RETURN
OF THE
WILD

RETURN
OF THE
WILD

The Future of Our Natural Lands

Edited by

TED KERASOTE

THE PEW WILDERNESS CENTER

ISLAND PRESS
Washington ▪ *Covelo* ▪ *London*

Island Press is a trademark of The Center for Resource Economics.

The Pew Wilderness Center, funded by a grant from The Pew Charitable Trusts,
is a project of Earthjustice.

Library of Congress Cataloging-in-Publication Data
Return of the wild : the future of our natural lands / edited by Ted Kerasote.
  p.   cm.
Includes bibliographical references (p.   ).
ISBN 1-55963-926-1 (hardcover : alk. paper) — ISBN 1-55963-927-X (pbk. : alk. paper)
1. Wilderness areas—United States. 2. Nature conservation—United States.
I. Kerasote, Ted.
QH76 .R48 2001
333.78′2′0973—dc21   2001004279

British Library Cataloguing-in-Publication Data available.

Printed on recycled, acid-free paper.♻
Manufactured in the United States of America
2 4 6 8 10 9 7 5 3 1

We must remember always that the essential quality
of the wilderness is wildness.

— HOWARD ZAHNISER
FATHER OF THE 1964 WILDERNESS ACT

Contents

RETURN

OF THE

WILD

Introduction
Ted Kerasote

From my house in northwestern Wyoming I can see three designated wilderness areas: the Gros Ventre, the Jedediah Smith, and the Teton. A short walk up one of the nearby foothills reveals a more awe-inspiring sweep of wild country: the Wind River Range to the south, Yellowstone National Park to the north, and—if the sky is very clear—the Sawtooth Mountains as well as the sprawling River of No Return Wilderness to the west, in central Idaho.

Even if clouds swirl over the river bottoms, obscuring these grand vistas, the sense of wildness in this place is oceanic, the tidal migrations of wildlife—pronghorn, elk, and bison, wolf, grizzly, and lion, eagle, swan, and crane—adding to my sense that this world is out of another age.

I am not alone in appreciating wild landscapes. During the last decade, counties adjacent to such places experienced enormous population growth, far outstripping other regions in the United States. This is not merely a reflection of human numbers increasing and people needing new places to live.

As mass marketing continues to turn our cities, houses, appliances, and work stations into dull clones of each other, individuality—in ourselves and in our surroundings—becomes ever more precious. We hunger for character and difference, and nowhere are these qualities more apparent than in the nation's wildlands: the misty hardwood forests of the Adirondacks, the red rock canyons of Utah, the conifer-banked rivers of the Pacific Northwest, the watery prairie of the Everglades. Each has its own inimitable personality—a mixture of sunlight, aroma, and texture underfoot—that we feel like an embrace and often return with tenderness.

Few of us are so civilized as to be immune to this exchange. Indeed, even the most dedicated of city dwellers have stood on my porch, stunned by the silence, the space, and the purity—on their faces a melting regard, their defenses lowered in the presence of the newborn creation.

Such sentiments with respect to our diminishing bank of wild country aren't new—they've been around since the early 1800s when the mountain men, first in the Appalachians and then in the Rockies, bemoaned civilization's destruction of the splendid innocence that had belonged to them, the Indians, and the buffalo alone. What is different at the beginning of the twenty-first century is that the recognition of this loss has become more widespread in our culture. Almost all of us have seen a beloved landscape disappear under houses, pavement, and retail space, and our personal view of how our civilization works mirrors what the geographers tell us.

According to the World Resources Institute, 30 percent of the planet has already been converted to agriculture and urban development, and in less than a century it's predicted that 30 percent more will be cut, plowed, and paved. This will leave only about one-third of Earth's land area in a natural state by 2100,[1] and one-tenth of these wild places lie within Greenland, northern Canada, Antarctica, and alpine glaciers—they're virtually all ice. The fate of the remaining old-growth forests, unfenced grasslands, and deserts without powerlines is what's at stake. Today, only 9 percent of this undeveloped country worldwide has been set aside in some form of protected area[2]—reservoirs of biological diversity, sources of clean air and water, wild foods and new medicines, sinks for carbon in an age of global warming, and places for us to escape to nature.

Some of us, working and living in cities, may dismiss the importance of these wild places, but the fact remains that civilization can't exist without the ecological services they provide. In addition, millions of us—urban, suburban, and rural dwellers alike—still find that our most idyllic and personally challenging moments happen in wilderness.

These issues are, in essence, the subject of *Return of the Wild*: why nature in its most untrammeled state is vitally important; what currently threatens wild country; and what can be done to conserve more of it. We at the Pew Wilderness Center are unabashedly narrow in our focus. We believe the fate of these lands will reflect the quality of all life on Earth during the next several hundred years. Consequently, we wish to help conservation organizations set aside more wild country. Some of it could be added to the 106-million-acre National Wilderness Preservation System, which confers permanent protection from any sort of development (roads, houses, logging, mining, energy exploration) and permits only nonmotorized recreation such as hiking, horsepacking, canoeing, fishing, skiing, and hunting. Other portions of the nation's roadless country could be given varying degrees of protection that insure open space and intact wildlife habitat.

While our society's consciousness about the importance of these wild places has become elevated, the on-the-ground efforts to conserve them has unfortunately remained the work of a few. As individuals and as a nation we still believe that if we recycle and buy more fuel-efficient cars we have done our bit as environmental stewards. Yet, if we don't conserve more of the still-undeveloped places of Earth, human life will become increasingly disconnected from its fellow animals and torn from its roots. Nature will continue to be dismantled and species will continue to wink out by the thousands. We humans will still exist, of course, but we'll be like potted trees in the foyers of great skyscrapers—alone and not part of a wider forest. Our efforts to recycle and use energy more efficiently must therefore be augmented with advocacy for the protection of wild spaces. *Return of the Wild* is intended to be the educational underpinning for that endeavor: a guide through the issues of the day, a history, a forum for debate, a source of information.

To this end, the first annual edition of *Return of the Wild* contains a new mapping of all the roadless areas in the United States larger than one thousand acres that remain on federal lands and which are not part of the National Wilderness Preservation System. The map was created by Pacific Biodiversity Institute in Winthrop, Washington and shows the nation to have 408.2 million acres of roadless lands that meet this definition. About 30 million of these acres are found in national parks and have some degree of protection. The remaining 380 million acres—153 million acres in Alaska, 227 million acres in the contiguous United States, and a few thousand acres in Hawaii, Puerto Rico, and the U.S. Virgin Islands—are fragmented to varying degrees. In Alaska millions of acres exist in vast untouched swaths; in the eastern part of the United States undeveloped tracts of land are highly segmented by roads. Nonetheless, these blocks of land, measuring from one to five thousand acres in size, are valuable to wildlife as well as for the maintenance of water and air quality. When forested, as they are in the eastern part of the nation, they can give recreators the sense that they are far from the noise and sights of the modern age. Eventually, portions of this roadless country could be included in the National Wilderness Preservation System. The latter is also shown on this fold-out map, on the reverse side of which is a sobering view of the entire road system of the lower forty-eight. Except for a scattering of blank wild spaces, the nation is almost full of corridors for motor vehicles.

At the beginning of the twentieth century a few far-seeing individuals recognized that this sort of development was imminent and that action had to be taken if some vestige of wild America was to be preserved. In the essay following the fold-out maps, "A Brief Illustrated History of Wilderness Time," Douglas W. Scott, the Pew Wilderness Center's policy director, describes how the American conservation movement was born, its key figures, and the events that led to the creation of a National Wilderness Preservation System in 1964.

Vine Deloria, Jr., a Native American scholar and the author of *Custer Died For Your Sins* and *God Is Red*, adds a cautionary note to this inspiring tale, reminding us that the very idea of wilderness is a European notion. What Europeans found to be an intimidating continent and tried to dismantle, and what later Americans tried to preserve in parks and wilderness areas, Native Americans saw as home. Deloria addresses why the two cultures had such different perceptions of landscape and how in the twenty-first century these differences still exist, coloring our views of how land should be treated.

Chris Madson, the editor of *Wyoming Wildlife*, closes Part I, "Our Wilderness Heritage," with a discussion of a little-recognized but important segment of Euro-American culture that has found more delight than terror in wild country and since the late 1800s has worked to conserve it: sport hunters and anglers. The first champions of protecting wild space, these two groups—now numbering 17 million hunters and 60 million anglers—could unite, Madson tells us, with environmental organizations to create a powerful voice for wildlands protection.

In Part II, "The Human Landscape," Thomas Michael Power, an economist at the University of Montana, lays to rest one of the oldest critiques of wilderness preservation. When there are mouths to feed and people who need jobs, goes this argument, how can we remove part of the nation's land base when it could be employed for gainful purposes—energy development, logging, and motorized recreation? His counter argument should interest chambers of commerce and county commissioners nationwide: wilderness areas are one of the driving forces of the new U.S. economy. Attracting light industry and a variety of information-age professionals, wilderness areas contribute markedly to a region's economic well-being.

But the very people who are driving new wildland economies—scientists, engineers, software developers, and computer and communication technicians—are also the sort of folks who want to build their dream homes on the edge of wilderness, alongside wildlife and spectacular views. Florence Williams, who has written about environmental issues for *The New York Times* and *Outside*, describes how these lovers of wild country can also be its enemies as they contribute to the subdivision of private property, often ranches, that are the buffer zones between towns and wild space. She also illustrates how these newcomers to the West can be a key component in forming partnerships with the old ranching economy to protect private lands.

Mike Matz, the director of the Pew Wilderness Center, then gives his commentary on the current legislative battles over wildlands protection and the key players in both Congress and conservation organizations. Why does one roadless area become a designated wilderness while another languishes unprotected? Matz leads us through the swirling political realliances in the changing Democratic and Republican parties.

Realliance is also the subject of Steven Bouma-Prediger's essay. An associate professor of religion at Michigan's Hope College, he describes the Biblical justification for humans abandoning their role of domination vis-à-vis nature and adopting one of compassionate stewardship. In a short accompanying essay, he outlines the movers and shakers of the Christian environmental movement and how they are becoming a force to be reckoned with in wildlands preservation. Suellen Lowry, an attorney who has facilitated dialogue between religious and environmental groups, offers a practical sidelight to Bouma-Prediger's essay by describing how secular environmentalists can join with religious groups to form powerful coalitions that can affect public policy.

In Part III, "Wildlife and Wildlands," Jack Turner, long-time adventurer, philosopher, and author of *The Abstract Wild*, examines one of the least known and most frightening threats to wildness: genetic engineering and transgenic species. He concludes that genetically modified forests and fish, to take just two examples, have the ability to alter nature completely. To save the wild, says Turner, we must stop managing it.

Michael Soulé, one of the founders of the Society of Conservation Biology and The Wildlands Project, answers that intensively managing wilderness areas—albeit with care—is both necessary and appropriate in a time when nature is being overwhelmed by unprecedented threats. His analysis of how that management could be done to restore the ecological health of wild areas will hearten and anger people on both sides of the debate.

Hal Herring, whose work has appeared in *The Atlantic Monthly* and *High Country News*, follows both arguments with a look at how game ranching is diminishing the wildness of elk, for many North Americans the symbol of remote, undeveloped country. Such manipulation of wildlife, Herring concludes, is but a symptom of our culture's entrenched belief that nature can be improved through tinkering.

Todd Wilkinson, an environmental correspondent for the *Christian Science Monitor*, closes this section with a report on what might be called "tinkering for a noble end" and how the unlikeliest of partners—central Idaho loggers, environmentalists, and sportsmen—are trying to restore grizzly bears to the second largest roadless area in the contiguous U.S. Even though Interior Secretary Gale Norton recently shelved the plan, it remains a blueprint for restoring endangered species.

In Part IV, "Heart of the Wild," Richard Nelson, an Alaskan anthropologist and author of *The Island Within* and *Make Prayers To the Raven*, ends the book with a vivid personal essay about immersing ourselves more fully in wild environments near our homes. For Nelson, this immersion includes gathering edible plants and fishing and hunting for food—in other words, making the wild a literal part of his being.

For those unfamiliar with the 1964 Wilderness Act, one of the most eloquent pieces of legislation to emerge from the U.S. Congress, the text of the Act is included in the appendix. The Act allows unprotected wildlands to be considered for inclusion in the National Wilderness Preservation System, and citizens in many states continue to propose new candidates.

The optimistic message of *Return of the Wild* is that much wilderness remains to be savored in the United States. On the other hand, 80 percent of those wildlands remain unprotected, open space continues to be developed at the rate of about one million acres per year, and the energy, logging, and mining industries have found an open door to the nation's public lands via a conservative Republican administration. Ironically, that administration is trying to reverse the conservation legacy of its progressive Republican forefathers—men like Theodore Roosevelt, who set aside millions of acres of land from development, and Richard Nixon, who signed the Endangered Species Act into law.

We will need energy and focus if these benchmarks of wildlands and wildlife protection are to be maintained and expanded. We will also need to re-envision how our wildlands, chopped up as they have been into ever smaller segments by roads, can be reconnected to each other so as to restore their ecological integrity. To address these issues, the first annual edition of *Return of the Wild* will be followed in 2002 with a volume that describes how Canadian and Mexican wild country could be joined to that in the United States to form a continent-wide wilderness preservation system.

Given the estimated growth of the United States' population from 280 to 400 million people by 2050, and the attendant houses, agriculture, and roads that will be needed to support such a society, some of us may grow pessimistic that such a wilderness preservation system can ever be achieved. The language sprinkled throughout this book—*a precious legacy, the last, the best, the final home of freedom*—may reinforce the gloomy sentiment that we're at the end of a journey: an advanced civilization saving a treasured relic. This is one way to conceive of saving wild space, but it's not the way I hope *Return of the Wild* is interpreted. I prefer that it envisions a beginning.

Native Americans inhabited this continent at least twenty thousand years before creating the landscape many of us now refer to as having achieved a balance between nature and humans. In fact, the new migrants to North America—armed with spears and cooperative hunting strategies—helped in driving the continent's naïve wildlife to extinction, animals like mammoths, western camels, and giant short-faced bears. It took the descendants of these first immigrants to North America thousands of years to learn how to live on the continent without destroying its

7

wildlife, forests, and grasslands—to become, as Vine Deloria, Jr. says, truly indigenous.

We in the twenty-first century do not have that sort of luxury—our technology is exceedingly powerful and there are so many of us. Yet, the process must be similar: we too must become indigenous inhabitants of our home places, creating a civilization that complements wildness rather than destroys it. Our learning curve must necessarily be steep, and we must find compassionate ways to control human population growth, but people of good faith have accomplished other great tasks in a short time.

There is no better way to begin this task than by walking away from the roads for a few hours…for a few days…for a few weeks if we are so fortunate. In the silence we may discover, as John Muir so well observed, that wildness is a necessity and that returning to it is going home.[3]

PART I

OUR WILDERNESS HERITAGE

I am glad I shall never be young
without wild country to be young in.
Of what avail are forty freedoms
without a blank spot on the map?

— ALDO LEOPOLD

How the Wildlands Map Was Produced
John McComb

～ The map was produced by Pacific Biodiversity Institute, using geographic information system (GIS) technology to analyze roads on federally owned land in the four major land management agencies. These agencies are the Forest Service, National Park Service, Fish and Wildlife Service, and the Bureau of Land Management. Jeep trails and similar tracks are not considered roads.

The National Wilderness Preservation System (NWPS) is depicted as it generally existed at the end of 1999 along with ten areas in Nevada that were added to the NWPS in 2000. Other smaller areas added in 2000 are not included. Several states have wilderness systems that are also shown on the map; this data is incomplete especially in Alaska and Maryland. Not included are Indian reservations and military lands. The State of Alaska is atypical both in the total amount of designated wilderness and in the potential for adding large new areas to the NWPS.

The U.S. Road Network map displays the road data that was used in this analysis. Over seven million miles of roads have fragmented the once primeval landscape that European explorers found more than four hundred years ago. A new road may occupy only a relatively small area, but it can have a large impact by fragmenting a previously contiguous block of potential wilderness. The accompanying chart and bar graph include both wilderness and non-wilderness roadless areas on public lands.

CAVEATS

This analysis looked only at ownership and roads and did not consider other landscape qualities that may affect wilderness suitability, such as an area having been logged by helicopter. A number of factors may result in reduced accuracy in the maps, including poor source data from the agencies and how water areas are counted. In many states wilderness advocates have completed a more detailed inventory and in some cases detailed wilderness proposals.

For more detail on any area visit the Pew Wilderness Center website at *www.pewwildernesscenter.org*. Maps in Adobe Acrobat PDF format at a scale of 1:1,000,000 (Alaska 1:2,500,000) are available on this site. For the western United States and some other states there is a separate map for each state. Smaller states have been grouped together.

ALASKA

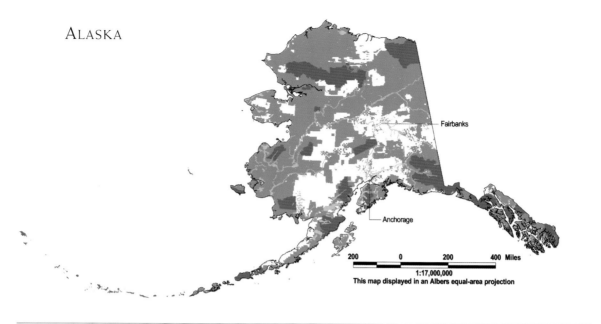

Fairbanks

Anchorage

| 200 | 0 | 200 | 400 | Miles |

1:17,000,000
This map displayed in an Albers equal-area projection

HAWAII

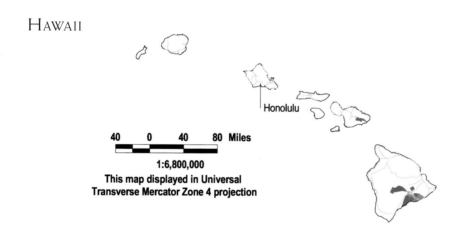

Honolulu

| 40 | 0 | 40 | 80 | Miles |

1:6,800,000
This map displayed in Universal
Transverse Mercator Zone 4 projection

PUERTO RICO & U.S. VIRGIN ISLANDS

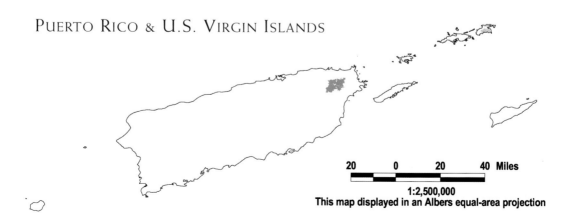

| 20 | 0 | 20 | 40 | Miles |

1:2,500,000
This map displayed in an Albers equal-area projection

This map is based on our current compilation of the most
comprehensive road data for the United States. It represents
road data collected from various federal and state agencies.
Some smaller roads are not shown on this map.

Pacific Biodiversity Institute
www.pacificbio.org
2001

The U.S. Road Network

200   0   200   400 Miles

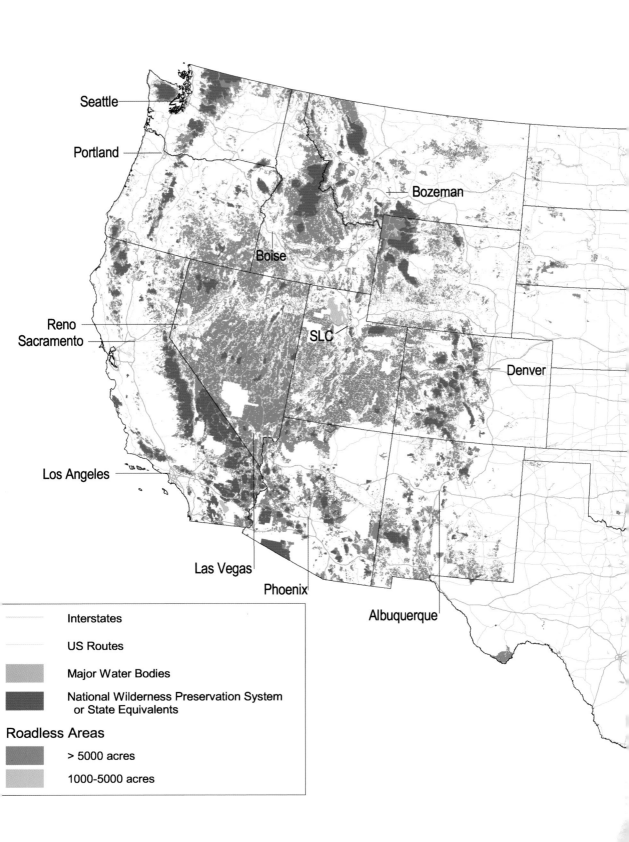

Seattle

Portland

Bozeman

Boise

Reno
Sacramento

SLC

Denver

Los Angeles

Las Vegas

Phoenix

Albuquerque

Interstates

US Routes

Major Water Bodies

National Wilderness Preservation System
    or State Equivalents

Roadless Areas

> 5000 acres

1000-5000 acres

THE WILDLANDS
OF THE UNITED STATES

Minneapolis

Boston

Chicago

New York

Washington

Charleston

Atlanta

Dallas

New Orleans

Jackson

Miami

200   0   200   400   **Miles**

1:15,000,000
This map displayed in an Albers equal-area projection

Pacific Biodiversity Institute
www.pacificbio.org
2001

Wilderness and Non-wilderness Roadless Areas

United States	Agency Total	NWPS*	% Land[1]	Non-wilderness Roadless	% Land[1]
Bureau of Land Management	264,174,745	6,228,152	0.3%	211,711,639	9.3%
Forest Service	192,046,672	34,590,393	1.5%	107,949,770	4.8%
Fish and Wildlife Service	93,628,302	20,686,134	0.9%	60,553,945	2.7%
National Park Service	77,937,494	44,046,459	1.9%	28,004,109	1.2%
Total	627,787,213	105,551,138	4.6%	408,229,463	18.0%

Except Alaska (49 states, Puerto Rico, and U.S. Virgin Islands)	Agency Total	NWPS*	% Land[2]	Non-wilderness Roadless	% Land[2]
Bureau of Land Management	177,723,471	6,228,152	0.3%	128,900,962	6.8%
Forest Service	170,048,058	28,838,172	1.5%	93,123,173	4.9%
Fish and Wildlife Service	16,646,826	2,009,222	0.1%	4,740,999	0.2%
National Park Service	26,852,459	10,293,376	0.5%	11,280,817	0.6%
Total	391,270,814	47,368,922	2.5%	238,045,951	12.5%

Alaska	Agency Total	NWPS*	% Land[3]	Non-wilderness Roadless	% Land[3]
Bureau of Land Management	86,451,274	–	0.0%	82,810,677	22.7%
Forest Service	21,998,614	5,752,221	1.6%	14,859,785	4.1%
Fish and Wildlife Service	76,981,476	18,676,912	5.1%	55,812,947	15.3%
National Park Service	51,085,035	33,753,083	9.2%	16,728,074	4.6%
Total	236,516,399	58,182,216	15.9%	170,211,483	46.6%

Agency and wilderness acreages are from agency data.
Non-wilderness roadless acreage is from Pacific Biodiversity Institute mapping study.
* NWPS – National Wilderness Preservation System
[1] % Land is of the total land area of the United States – 2,271,343,000 acres.
[2] % Land is of the total land area of the 49 states, Puerto Rico, U.S. Virgin Islands excluding Alaska.
[3] % Land is of the total land area of Alaska.

Size Distribution for Wildlands on Public Land
(includes both wilderness and non-wilderness roadless)

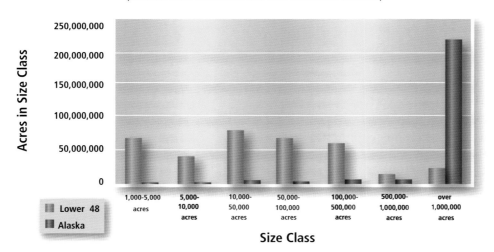

A Brief History of Wilderness Time

Douglas W. Scott

1830–1900

"How near to good
is what is *wild!*"
—Henry David Thoreau

"Wilderness," said Aldo Leopold, "is the raw material out of which man has hammered the artifact called civilization." For more than two centuries North American settlers concentrated on the hammering, giving little attention to saving the fast-disappearing raw material. It was only in the middle of the nineteenth century that essayists, poets, and artists evolved a vision of wild nature that made it valuable in its own right. Their work in turn fired the broader public imagination. Wild nature became a cherished part of what was felt to be unique about North America.

Few individuals were more influential in creating this popular vision than Henry David Thoreau, the recluse of Walden Pond who extolled the value of "the tonic of wildness" and asserted that "life would stagnate if it were not for the unexplored forests and meadows which surround it." With the close of the frontier at the end of the nineteenth century, more and more Americans found resonance with Thoreau's

Canvases like "Yosemite Valley," painted by Albert Bierstadt in 1868, helped Americans visualize the beauty of nature. Such paintings were exhibited in eastern cities and drew tens of thousands of viewers.

assertion that "we can never have enough of nature."[1]

1903

"Wildness is a necessity."

—JOHN MUIR

President Theodore Roosevelt and John Muir on the rim of Yosemite Valley, 1903. Muir, ever the propagandist for wilderness, hoped to "do some forest good in freely talking around the campfire." The two camped alone, reveling in the four inches of snow that fell on them overnight. T. R. returned to Washington, D.C. exulting that it was "the grandest day of my life!"[2]

Even as Americans were coming to value their unique heritage of wilderness, they were also confronted by an urgent historical reality: unspoiled country was fast disappearing from the landscape. In no small measure, the desire to preserve the remaining wilderness arose in reaction to the rapidity with which primeval America was being conquered.

At the turn of the nineteenth century, John Muir became the great popularizer of what he called "going to the mountains." He was a virtual one-man publicity factory churning out articles and books that promoted wild forests and national parks as "fountains of life," affording "thousands of God's wild blessings" for those who would "climb the mountains and get their good tidings." The Sierra Club, which he helped found in 1892, today carries on Muir's work for wilderness.

Theodore Roosevelt, of whom Muir said, "I never before had so interesting, hearty, and manly a companion," was the first president with an activist agenda for conservation. He created a national system of refuges that afforded havens to wildlife, and he liberally employed the 1906 Antiquities Act, which empowers the President to set aside unique natural and historic areas on federal lands as "national monuments" (a practice presidents continue to this day). Roosevelt also worked to protect national forests. As he told

Congress in 1901, the national forests "should afford perpetual protection to the native fauna and flora, safe havens of refuge to our rapidly diminishing wild animals of the larger kinds, and free camping grounds for the ever-increasing numbers of the men and women who have learned to find rest, health, and recreation in the splendid forests and flower-clad meadows of our mountains."[3]

1924

"We should keep here and there...some bit of wilderness frontier."
—WILLIAM B. GREELEY, CHIEF FORESTER

In 1872, Congress established the world's first national park—Yellowstone. More soon followed. But as tourism boomed, the parks came under constant pressure from boosters who wanted more roads, hotels, marinas, and car campgrounds. This drumbeat for development, not to mention the National Park Service's dual mission—to leave its landscape "unimpaired for for the enjoyment of future generations" while simultaneously opening its gates for the pleasure of the American people—made early conservationists skeptical of the Service's ability to preserve wild places in the long run.

Therefore, it wasn't all that surprising that the first steps for preserving wilderness occurred on other federal lands. The young, Yale-educated Aldo Leopold was the driving force behind protecting roadless areas in the national forests. In 1924, he convinced his superiors to set aside 574,000 acres of New Mexico's Gila National Forest as a wilderness area. With Leopold's advocacy, the idea of protecting more federal wilderness

University of Wisconsin, Madison Archives

Aldo Leopold—forester, wildlife biologist, essayist, and philosopher—on a 1924 canoe trip in what is now the Boundary Waters Canoe Area Wilderness of northern Minnesota. "The number of adventures awaiting us in this blessed country seems without end," he wrote in his journal. In 1949, Leopold's A Sand County Almanac *was published. It remains in print today, providing the ethical underpinning of modern environmentalism.*[4]

areas from road-building and logging spread. In 1929, the Forest Service instituted the first national wilderness policy, designating 14 million acres of the western states as "primitive areas."[5]

1939

"The harsh environment of untrammeled expanses..."
—ROBERT MARSHALL

In the 1930s, Robert Marshall joined Leopold as one of the leaders of the wilderness preservation movement. An energetic, independently wealthy young forester, Marshall had grown up amid the influence of the Adirondack Forest Preserve, designated "forever wild" by the New York State Constitution. An exuberant proselytizer for wilderness, who gathered acolytes from the Great Smokies to Alaska, Marshall saw wilderness as a place for men and women to test themselves against the elements; his hikes of thirty or more miles in a day were legendary. Decrying the loss of wildlands to the seemingly endless spread of roads, Marshall said that "it may be a better boast for Utah that she still retains the largest roadless area remaining in the United States than that she has more miles of State road than Arizona or Wyoming."

As a senior administrator of the Forest Service in the late 1930s, Marshall pressed to have "primitive areas" restudied and reclassified as "wilderness areas," which would have more lasting protection from logging and road-building. Like Leopold, Marshall advocated wilderness preservation not only on national forests, but also on the still-wild portions of other federal lands. To this end, he helped organize The Wilderness

University of California, Berkeley, Bancroft Library

Robert Marshall, legendary hiker and wilderness exponent, on Alaska's Mt. Doonerak in 1939. This 7.3 million-acre wilderness is now contained within Gates of the Arctic National Park. In his journal, Marshall wrote "no comfort, no security, no invention, no brilliant thought which the modern world had to offer could provide half the elation of twenty-four days in the little-explored, uninhabited world of the arctic wilderness."[6]

Society in 1935, but his sudden death in 1939, at age thirty-eight, robbed the conservation movement of one of its great champions.[7]

1950

"From the eternity of the past... into the eternity of the future..."
— HOWARD ZAHNISER

Addressing the still-tenuous position of America's wilderness areas in 1939, the Izaak Walton League's Kenneth Reid wrote, "There is no assurance that any or all of them [the designated wilderness areas] might not be abolished as they were created—by administrative decree. They exist by sufferance and administrative policy—not by law."

His were not idle fears. The pace of road building, logging and other development on the federal lands accelerated in the long economic boom after World War II. Meanwhile, there was no real national policy to preserve wilderness for all categories of federal lands; there wasn't even a practical definition of what a wilderness area was. Faced with the continuing loss of wild places, wilderness advocates yearned for a stronger, more dependable means of preserving wilderness in perpetuity.

Into the breach stepped Howard Zahniser, becoming the executive director of The Wilderness Society in 1945. He and the Society, along with the Sierra Club, the National Parks Association, and the National Wildlife Federation, made it their goal to secure a federal statute that would protect wilderness areas, particularly those that lay outside national parks and monuments. "Let us try to be done," Zahniser said, "with a

Photo by James Marshall, courtesy of Ed Zahniser

Howard Zahniser, one of the least known and most influential conservationists of the twentieth century, during a 1955 pack trip in Idaho's Selway-Bitterroot Primitive Area. A scholarly professional editor who became executive director of The Wilderness Society, Zahniser conceived and drafted the 1964 Wilderness Act and led the campaign for its enactment.

wilderness preservation program made up of a sequence of overlapping emergencies, threats, and defense campaigns!" What he envisioned instead was a positive program in which a nation-wide system of designated wilderness areas would have the firm protection of federal law. Given the concerted opposition from powerful logging, grazing, mining, energy and tourist industries, and their friends in Congress, it was a bold ambition.[8]

National Park Service Historic Photographic Collection, Harpers Ferry Center

Before they could launch the campaign for the Wilderness Act, advocates were confronted with a fundamental assault on the integrity of the National Park System and its wilderness: a proposed dam on the Green River, drowning Echo Park.

1955

"We'll hand them a tool they'll use
for the next hundred years."
— REP. WAYNE ASPINALL
(D-COLORADO)

Zahniser's campaign for wilderness legisla-tion was sidetracked in the early 1950s when western water development interests proposed a dam within the then obscure Dinosaur National Monument, located on the Utah-Colorado border. Because the Echo Park Dam—one element of a billion-dollar package of dams across the Colorado Plateau—would invade a unit of the National Park System, "Zahnie," David Brower of the Sierra Club, and Fred Smith of the National Parks Association declared it wrong in principle. Until the authorization to build this dam was removed, they vowed to block congressional approval of the entire billion-dollar regional water project.

The ensuing fight over the Echo Park Dam was the greatest conservation battle of the twentieth century. Thousand of Americans wrote to Congress opposing the dam, and hundreds of editorials were written denouncing this attack on America's national parks. In this epochal struggle, the modern U.S. environmental movement was forged. Both sides knew the stakes. The leading dam advocate in Congress, Representative Wayne Aspinall from Colorado, said, "If we let them knock out Echo Park Dam, we'll hand them a tool they'll use for the next hundred years."

Against all odds, the conservation groups won. By 1955, water developers agreed to drop Echo Park Dam so as to gain congressional approval for the remainder of their regional package. In winning, the conservationists established the political credibility and momentum to begin their campaign for a Wilderness Act. A week after the Echo Park victory, Zahniser began to draft the bill.[9]

Courtesy of National Park Service Historic Photographic Collection, Harpers Ferry Center; Photo by Abbie Rowe

The Rose Garden, September 3, 1964. President Lyndon B. Johnson has just signed the Wilderness Act and hands pens to Mardy Murie (left) and Alice Zahniser. Their husbands, Olaus and Howard, were key leaders of the conservation forces and had died during the long lobbying campaign. Secretary of the Interior Steward L. Udall leans over the president as pro-conservation members of Congress look on: Sen. Frank Church (D-Idaho) behind Murie; Wayne Aspinall (D-Colorado) behind Zahniser; Sen. Clinton P. Anderson (D-New Mexico) to the right of Aspinall; and Rep. John P. Saylor (R-Pennsylvania) above Udall's back.

1964

"Wilderness is the original sustainable development."

—HOLMES RALSTON III

The Wilderness Bill, still a very long shot, was introduced in Congress in 1956. Over the next eight years, it went through eighteen hearings, covering nearly three thousand pages of closely printed transcript. The bill attracted concerted opposition from the logging, mining, grazing, and water-development interests, from conservative western congressmen, and even from the Forest Service and the National Park Service, since neither agency wanted to give up its ability to develop wildlands. The break came when John F. Kennedy became president in 1961 and endorsed the bill.

Even with presidential support, struggles to overcome the opposition of conservative opponents took another four years. Led by progressive westerners, the Senate passed the Wilderness Bill in 1961, but it died in the House the following year. The Senate passed it again in 1963 by a vote of 73 to 12. With agreement on some final compromises, the House followed in 1964 by a vote of 373 to 1, and President Lyndon B. Johnson signed the Wilderness Act into law.

Howard Zahniser was not there to savor the moment—he had died four months before, days after testifying at the final congressional hearing. He, his colleagues, their congressional supporters, Presidents Kennedy and Johnson—and millions of citizens who urged Congress to act—had given an enduring gift to the American people.

The Act made it the national policy of the United States to preserve areas of wilderness on the federal lands, saying that "a wilderness, in contrast with those areas where man and his own works dominate the landscape, is hereby recognized as an area where the earth and its community of life are untrammeled by man, where man himself is a visitor who does not remain." The Act created a National Wilderness Preservation System (NWPS) and went on to define wilderness areas as those which retain their "primeval character and influence, without permanent improvements or human habitation." Wilderness also has "outstanding opportunities for solitude or a primitive and unconfined type of recreation." Fishing, hiking, camping, skiing, mountaineering, and horsepacking are all permitted in wilderness areas; motorized vehicles or machines are not. Hunting is permitted in those wilderness areas that lie outside of national parks.

Congress alone can designate wilderness areas and fix their boundaries. Most importantly, once an area is designated as wilderness, an Act of Congress is required to eliminate so much as an acre or alter a boundary. A prospective wilderness area must be five thousand acres or "of sufficient size as to make practicable its preservation and use in an unimpaired condition." The more than 640 wilderness areas designated through the year 2000 include areas as large as 9.6 million acres and as small as a five acre island.[10]

Photo by J.F. Carithers, courtesy of Ed Zahniser

Leaders of The Wilderness Society at their annual meeting to discuss strategy for preserving wildlands. This 1959 gathering took place at the edge of the Blue Range Primitive Area in Arizona. Left to right: James Marshall (older brother of Bob Marshall), Sigurd Olson, Olaus Murie, George Marshall (younger brother of Bob Marshall), Richard Leonard (also a Sierra Club leader), Harvey Broome, Ernest Griffith, Stewart M. Brandborg, Howard Zahniser, and Bob Cooney.

1964–1975
"A great liberating force in the conservation movement..."
—STEWART M. BRANDBORG

Only 9,130,000 acres of wilderness were initially included in the NWPS, but a ten-year study was mandated for remaining "primitive areas" as well as roadless regions within national parks and wildlife refuges. These studies resulted in recommendations to Congress to designate additional wilderness areas.

Having Congress vote on each addition to the

NWPS might seem a drawback since passing a bill through that body is often a daunting task. However, Zahniser's successor at The Wilderness Society, Stewart M. Brandborg, recognized in this challenge the hidden potential for "a great liberating force in the conservation movement." As citizens across the nation studied federal lands near their communities, attended public hearings, and actually wrote their own wilderness proposals, a grassroots conservation network was formed. In this way, the wilderness preservation movement expanded and built a decentralized power base. As a result of these citizens' efforts Congress has enacted more than one hundred wilderness laws since 1964, adding more than 96 million acres to the National Wilderness Preservation System.[11]

1964–1980

"Alaska wilderness areas
are truly this country's
crown jewels."
—President Jimmy Carter

The Wilderness Act itself designated no wilderness areas in Alaska, yet the federal wildlands there are as vast and unspoiled as any on Earth. In 1971, Congress responded to conservationists' requests, requiring a study of the state's unprotected wild places. In 1977, with these studies in hand, wilderness activists mobilized an enormous national campaign focused on Alaska.

Under the tireless leadership of President Jimmy Carter, the Alaska National Interest Lands Conservation Act became law in 1980. This single act more than doubled the acreage of protected wilderness in America.[13]

President Jimmy Carter and his wife, Rosalynn, relax on the tundra of the coastal plain of the Arctic National Wildlife Refuge with Alaska wilderness leader Debbie Miller and her daughter, Casey. On this 1990 trip, the Carters closely observed musk oxen and, at one point, were surrounded by eighty thousand caribou. It was, Carter recalled, "a profoundly humbling experience. We were reminded of our human dependence on the natural world."[12]

1970–2001

"The de facto wilderness is the
wilderness that waits in death row."
—David R. Brower

Photo by John McComb

*Sweet victory after a long
and difficult campaign.
One of the greatest
congressional champions
of protecting federal lands,
Rep. Morris K. Udall
(D-Arizona) replenishes the
champagne as grassroots
activists celebrate passage of
another wilderness bill in the
offices of the Sierra Club.*

The Wilderness Act did not require
that all roadless, undeveloped federal
lands be studied for possible designation as wilderness—vast roadless
portions of the national forests were left
out. Conservationists called these areas
"de facto wilderness." In fact, the Sierra
Club's Dave Brower quipped that these
lands had "been set aside by God" but
hadn't "yet been created by the Forest
Service." Tens of millions of acres of western
deserts and rangelands, managed by the federal
Bureau of Land Management, were also not
required to be studied.

But the Wilderness Act left the door open for
local citizens to take their own wilderness proposals directly to Congress. Faced with resistance
from some in the U.S. Forest Service to designating new wilderness areas, that is just what
citizens did for areas in Montana, Oregon,
Alabama, West Virginia, and other states. In the
early 1970s, Congress enacted a number of these
citizen wilderness proposals, and, in reaction, the
Forest Service undertook its own inventories of all
remaining roadless areas in the 1970s. Twelve
years after the Wilderness Act became law,
Congress also extended the wilderness study
requirement to the roadless areas administered by
the Bureau of Land Management. Eventually
many wilderness area proposals on national
forests were approved by Congress, and in 1994,
President Bill Clinton signed the California Desert

Protection Act, designating more than 4 million acres of BLM-administered lands as wilderness.

In 2001, President Clinton also signed a regulation protecting 58.5 million acres of roadless country administered by the Forest Service. In addition to setting aside these areas for scenery, wildlife habitat, and watersheds, sound economics had dictated their protection. The cost of building logging roads in these rugged areas exceeds the timber revenues returned to the U.S. Treasury. Their protection remains in doubt under President George W. Bush.[14]

TOWARD THE FUTURE

"In order to assure that an increasing population, accompanied by expanding settlement and growing mechanization, does not occupy and modify all areas within the United States..."
— OPENING WORDS OF
THE 1964 WILDERNESS ACT[15]

Were they alive today, the pioneers of wilderness preservation—Thoreau, Roosevelt, Muir, Leopold, Marshall, and Zahniser—would scarcely believe the progress made for wilderness protection. The National Wilderness Preservation System encompasses 105,800,000 acres (4.6 percent of the U.S. land base) in more than 640 units in forty-four states.

Behind the official designation of each of these wilderness areas lies an intense human story of the citizen advocates, agency officials, and members of Congress whose collective efforts helped to preserve these places with the strongest tool we have—a statutory law enacted by the Congress

Colorado Springs in front of Pikes Peak circa 1890 and again in 2000.

of the United States.

Millions of additional acres of federal lands—in national parks, in national wildlife refuges, on lands administered by the Bureau of Land Management, and in national forests—remain wild and roadless, eminently suitable for congressional protection. Citizen groups in virtually every state are working toward that end. Their success or failure will determine the face of the American landscape in the coming centuries.

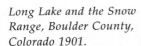

Long Lake and the Snow Range, Boulder County, Colorado 1901.

Photo by William Henry Jackson, courtesy of Colorado Historical Society

The same view, now preserved as the Indian Peaks Wilderness.

Photo by John Fielder

Editor's Note: Thank you to John Fielder
for the contribution of his photographs.

American Indians and the Wilderness

Vine Deloria, Jr.

∾ Every group of humans has at one time or another confronted a wilderness situation: a bewildering place where it appears that no one has gone before, where we feel the loneliness of nature the most. How different groups have resolved this anxiety-provoking interaction is fascinating. Some people have become one with the creatures of the land; others have imposed their own sense of order on what seemed an alien landscape, calling it a wilderness.

For the first European colonists the Atlantic seacoast was just such a place. Overwhelmed by forests and animals, they called the New World "howling," "dismal," and "terrible."[1] Later their grandchildren viewed the Kentucky bluegrass country in the same terms. People moving west to Oregon and California in the 1830s and 1840s cursed the mid-continent prairies as the Great American Desert. Walking for months in an endless sea of grass, they bemoaned their lot as they saw only wilderness around them. Subsequent explorations of the West would label the Great Basin Salt Flats, the Sonoran Desert of New Mexico and Arizona, the high desert of eastern Oregon, and the dreaded Death Valley of California as malevolent wildernesses with virtually no redeeming qualities.

American Indians also explored this country when it was new to them; however, as a rule, they saw it in different terms. If they exhibited the same uncertainty and fears as those displayed by the first European colonists then the experience—and its resolution—are now hidden in the mists of pre-history. I would doubt, though, that any Indian group experienced new lands in the way European settlers did.

Luther Standing Bear, a Brule Sioux chief from the Rosebud Reservation in South Dakota, gives one of the best explanations of the difference between the two cultures. "We did not think of the great open plains, the beautiful rolling hills, and winding streams with tangled growth, as 'wild,'" he writes in *Land of the Spotted Eagle*. "Only to the white man was nature a 'wilderness' and only to him was the land 'infested' with 'wild' animals and 'savage' people. To us it was tame. Earth was bountiful and we were surrounded with the blessings of the Great Mystery."[2]

Familiarity with the landscape and the understanding that everything was created by or dependent upon a higher power encouraged Indians to think of a homogeneity of life which ensured that nothing could be without meaning. That their lands were part of a larger cosmos was an important consideration in the Indian approach to nature. Hence no place could be what Europeans called a wilderness. It could only be part of a seamless web of life.

Yet white men also believed that everything was created by or dependent upon a higher power. Why then couldn't they live without radically harming the lands they settled in North America? Certainly part of the blame rests with the Christian doctrine of creation. Harvey Cox, a popular Protestant theologian, articulates rather precisely this attitude, derived from Genesis: "Just after his creation man is given the crucial responsibility of naming the animals. He is their master and commander. It is his task to subdue the earth."[3] It is this attitude that has been adopted wholeheartedly by Western peoples in their economic exploitation of the earth. The creation becomes a mere object when this view is carried to its logical conclusion— a result directly opposite from that of the Indian religions, which see a unity in nature, a bountiful earth where all things and experiences play a role.

Indeed, the task of the tribal religion, if such a religion can be said to have a task, is to determine the proper relationship that the people of the tribe must have with other living things and to develop the self-discipline within the tribal community so that man acts harmoniously with other creatures. The world that an Indian experiences is dominated by the presence of power, the manifestation of life energies, the whole life-flow of a creation. Recognition that human beings hold an important place in such a creation is tempered by the thought that they are dependent on everything in creation for their existence. There is not, therefore, the determined mission to subdue the earth and its living things. Instead the awareness of the meaning of life comes from observing how the various living things appear to mesh, providing a whole tapestry. There is no sense of conquest and no sense of dread.

Again, Standing Bear elaborates well on this idea by contrasting the manner in which Indians and non-Indians treated land. "Wherever forests have not been mowed down; wherever the animal is recessed in their quiet protection; wherever the earth is not bereft of four-footed life—that to him [the white man] is an 'unbroken wilderness.' But since for the Lakota there was no wilderness; since nature was not dangerous but hospitable; not forbidding but friendly, Lakota philosophy was healthy—free from fear and dogmatism."[4] Fear of the unknown, I would therefore suggest, constitutes a great part of the white man's perception of wilderness.

The white man, according to Standing Bear's characterization, stripped away various forms of life, plant, and animal, from the land and shattered the biotic unity that once existed there. Trees grow haphazardly, or so it seems; animals go their own way without the commands of humans. The need to destroy the living creatures of a landscape when viewed from the white man's perspective is simply that of bringing order to a land without perceived order so as to diminish his fear and increase his control.

The basic question that then emerges when comparing the two cultures is why has European culture been fearful of nature and Indian culture allowed itself to conform to nature's rhythms?

Standing Bear believed that the white man was a relative newcomer on the North American earth. "He is too far removed from its formative processes. The roots of the tree of his life have not yet grasped the rock and soil. The white man is still troubled with primitive fears; he still has in his consciousness the perils of this frontier continent, some of its fastnesses not yet having yielded to his questing footsteps and inquiring eyes. He shudders still with the memory of the loss of his forefathers upon its scorching deserts and forbidding mountain-tops."[5] The lack of roots coupled with the memories of hardships and the experience of traveling through or living in a wilderness, eliminated the chance of becoming indigenous. It is not difficult to see that the westward movement of white society, in which each generation leapfrogged forward perhaps as much as a thousand miles, prohibited the creation of the necessary psychic roots that would have given people confidence in confronting new landscapes.

There were, to be sure, many whites who lived successfully in wilderness areas. No one can deny that explorers, mountain men, and traders moved far into lands regarded as utter wilderness and lived successful lives. In almost every instance, however, they adopted Indian ways and emulated Indian techniques for living on the land.

These ways involved a people's surrendering themselves to the existence and purpose of the larger cosmos as represented in the earth. Standing Bear noted, "Men must be born and reborn to belong. Their bodies must be formed of the dust of their forefathers' bones."[6] We might call being born and reborn becoming indigenous. To operate on an entire culture that process would require generations of people submitting themselves to the rhythms of the land without attempting to make any radical changes to it.

How different was the pattern followed by white civilization! Boundary lines were drawn, rights were allocated, and groups of people were arbitrarily organized as political bodies simply because they fit within the statehood lines drawn by people thousands of miles away. But everything done in this respect only created an artificial way of life imposed upon a land bereft of its original biota and now subject to erosion rather than renewal. Clearly the idea of the ordered Garden, with man as the chief steward, controlled the white man's perceptions of landscape.

Wildlife could only flee from his advancing line of settlement, a phenomenon noted by Alexis de Tocqueville in his commentary on American institutions, *Democracy in America*. His words give us insight into the nature of a people's relationship to land. "As soon as a European settlement forms in the neighborhood of territory occupied by the Indians wild game takes fright. Thousands of savages wandering in the forest without fixed dwelling did not disturb it; but as soon as the continuous noise of European labor is heard in the vicinity, it begins to flee and retreat toward the west, where some instinct teaches it that it will find limitless wilderness."[7]

De Tocqueville emphasized this point frequently as if it had a mystical dimension that had not been perceived before. "A few European families," he observed, "occupying widely separated points succeed in chasing all the wild animals forever from the whole region stretching between these points."[8] The animals, of course, were wild only in the sense that they were *free*. De Tocqueville inquired of people why this phenomenon occurred and reported, "I have been assured that this effect of the approach of the white men is often felt at two hundred leagues from their frontier."[9]

Standing Bear, hardly a reader of de Tocqueville, made the same observations and drew the same conclusions concerning the effect of the white man on animals. "Between him and the animal there is no rapport and they have learned to flee from his approach, for they cannot live on the same ground."[10] How did animals become aware of the approach of civilized man two hundred leagues away?

Some of their awareness of civilization must have had to do with the noise settlements created. But Indian villages—full of chatter, babies crying, and dogs barking—must have been noisy places as well. The more important difference lies in how the two cultures conceived of and treated animals. In short, American Indians believed that the other entities of the natural world were also sentient, conscious beings. "Everything was possessed of personality, only differing with us in form," says Standing Bear. "Knowledge was inherent in all things. The world was a library and its books were the stones, leaves, grass, brooks, and the birds and animals that shared, alike with us, the storms and blessings of earth."[11]

If a people truly believed that the other entities of the natural world were also sentient beings—if they had experienced that vitality in other creatures over many generations—how could they conceive of the land in which these creatures lived as "wilderness"? There would not be a location or tract of land that would not be populated with some kinds of living beings who had much knowledge to share. Everything might be new but it could not invoke the feeling of disorientation that wilderness implied for early European settlers.

Still the question nags—how did such a view of the world evolve? To answer it let us examine how a band of Indians might have moved into a location where they had no knowledge of the landscape and might have found it puzzling and possibly malevolent. Someone in the band would have to discover the area and report back to the people what they had seen—an unknown or at least an unexplored bit of geography. They would explore this country gradually, moving as far as they deemed prudent with each venture. They would remember the most unique features of the landscape—the sources of water and the river systems, the birds and animals they encountered—and make comparisons with creatures they had known. People would discuss the new area and draw upon previous experiences with familiar adjacent lands to draw the proper analogies.

Most important for many tribes would be the configuration of the stars that appeared to be directly above at certain critical times: dawn, dusk, and during the phases of the moon. Some people would note when certain known stars rose and when they disappeared into notches and behind high points of the horizon. Most Indians believed that heaven and earth were inseparably joined, and that what happens above is reflected in what happens below and vice versa. So the stars could provide a map of the landscape that had not yet been explored. And once the star relationships were understood and memorized and had become familiar to people, future exploration of the area was not difficult. People would not get lost in the new lands because they could orient themselves with landmarks and with star patterns.

In addition, these newcomers would be particularly interested in trying to orient themselves to the behavior of the living beings already present on the land, trying to model their actions after the animals they saw around them and by so doing adapt themselves to the new landscape. In this respect, we see the Indian belief that humans are not the highest product of the creation or even of evolution. Instead, they fall short in many ways. They're not as fast as the four leggeds, not able to fly like the birds, not as keen of eye as the hawks and eagles, not as strong as the buffalo and bear. Because we were the last species created, according to Indian beliefs, humans were less well endowed than nonhuman animals. In the eyes of many tribes, we were the junior members of the land. Consequently, we needed to develop alliances with other beings that had been here longer and that possessed better physical attributes and had more wisdom than we.

Here we see how acutely the American Indian and the European settler differed. The Indian did not believe that this new land should be changed to fit familiar patterns of settlement and use that were known in the Old World. Rather, entering a new world, they would try to discern how to be as successful as the bird and animal peoples already living there.

The varied relationships of Indians with the bears of North America offer a very good example of how this learning was accomplished. If we take the wisdom of the bear possessed by the Indian tribes from the Great

Lakes to the Pacific Northwest, we find that their specific knowledge varies considerably. For some tribes, bear is a healer, for others a good guide to edible foods, for others he is influential in dreams and for still others he is a prophetic figure. Different landscapes would have created different behaving bears. Likewise, different landscapes would create different people—if, that is, they are careful observers of the life around them. Wilderness exists for humans only if we insist on refusing to accommodate to the land.

The Sioux offer us another example of accommodation. They were once woodlands people and subsequently moved onto the Great Plains. They realized that they knew nothing of how to live in that land of endless horizons so they watched the animals and adopted their ways. Moving through the woods meant following the trails of woodlands animals. The Great Plains had a few well-worn trails where the buffalo went, but in most places there were no good directions on how to travel. So the Sioux watched how the buffalo grazed and modeled their travel according to the buffalo's pattern.

A few formidable individuals moved ahead of the rest of the herd, acting as scouts to guide the herd as it grazed and migrated. On the flanks of the herd were other animals that could protect the herd from predators while the cows and the calves occupied the center position. Behind everyone would come the older, but still capable bulls to protect the herd from predators attacking from the rear. A Sioux village moving over the plains would use the same kind of organization with scouts, warriors on the side, the main body of women and children in the center and capable men at the rear to prevent an ambush.

The Sioux elders also say that they didn't know how to provide shelter for themselves on the Great Plains; one day, however, some boys were playing on a riverbank and a cottonwood tree told them how to make a good shelter by wrapping their leaves around a tripod of sticks. Thus was the tipi born with human adaptations of many more poles to support the basic structure. The people also observed that where they found turtles, they always found water, so they were especially alert to locate turtles whenever they went to an unfamiliar land. Consider as well the cricket called *ptepazopi* that used to live on top of the buttes on the plains and had little antennas on its head. The Indian scouts discovered that those antennas were always pointing in the direction of the nearest buffalo herd whether it was nearby or many miles away. We can guess that the crickets felt the vibrations of many buffalo grazing and running and therefore turned their antennas toward the herd to identify the direction the herd was moving. This discovery must have taken uncounted years of very careful observation of both buffalo and bugs to draw the conclusion that their behavior was related.

If a group of people found themselves on a strange landscape and lacked knowledge of how to feed themselves, this method of gaining information from other living beings—either in conversations with them or by observing them on the proper way to survive—meant that no tract of land could remain in the wilderness state for long before people became acclimated to it. A land could not be a wilderness if the people formed relationships with the creatures already present there and tried to use their methods of living. Early white explorers also learned to listen to birdcalls, watch grazing animals, and observe the kinds of food that certain animals ate. For them, too, the country was not a wilderness but home.

For Indians, it was also a home with some very special geographical features: Bear Butte in the Black Hills, the Devils Tower in Wyoming, the volcanoes in the Pacific Northwest, the mountains of Arizona, and the deserts of the Southwest and Great Basin. All of them became spiritual places for American Indians. Preliminary acquaintance with the areas would imprint their unique features on the minds of the people and they would conclude that these features had some special significance. So ceremonies and vision quests would be performed at these sites in an effort to discover what significance they held. The experiences that people had at certain locations became part of tribal knowledge and were passed down as special information for the future.

As people became familiar with a new landscape, they remembered events that had happened in certain locations, and they began to create a mental map of the region that reflected their experiences. Some prominent person might have been buried at one location, good ground potatoes might be found at other locations, a site where a band had divided into two groups when it became too large might be remembered, or a place where a great battle was fought might be marked. What might have been a wilderness quickly became a revered part of a people's memory. The Sioux can still point out places where century-old Sun Dances were held because the locations of the altar and posts have remained barren and without grass since the dance was originally performed.

Allusions to significant locations in previously occupied regions were also made and gradually these new lands took on the personalities of the original ones. For example, some members of the Five Civilized Tribes, who had once located a site where their people had emerged from the underworld in the Southeast, later designated a new site for the same tradition in Oklahoma after they had moved there. And the Sioux used to take their children to the White River in South Dakota, which had sites that released continuous white smoke from burning and buried lignite deposits. They used these locations to teach about the long-gone past when the elders said they were living in a hot steamy place with many volcanoes, presumably some part of Central America. These activities were designed to provide such a close relationship with the land that there would never be a sense of alienation or disorientation that might produce the idea of wilderness.

One question still remains: How did the Indians maintain themselves on the land without exhausting its resources? Each region had its own cycles of growth and maturity, and people tried to learn these cycles and behave in accordance with their rhythms. The Tohono O'odham of southern Arizona used star constellations to mark out the seasons for growth and harvest of desert plants. Thus when a certain constellation was overhead it was time to allow the animals to harvest the fruits. After the animals and birds had taken their share, humans could harvest for a short while and then the plants had to be left alone. In this fashion the people were guaranteed that there would always be a good supply of food.

Some tribes also obeyed spiritual directives in the use of their lands. The Iroquois followed the dictates of the Three Mothers—corn, beans, and squash—and always planted them together since it was believed they were compatible spirits. Ecologically, they are. Interplanting beans and corn provides some residual nitrogen, and both beans and corn reduce weeds by providing ground cover and shade. Even then, however, lands would become overburdened by continual planting, and villages might rotate to several locations within a described area, allowing some fields to lay fallow so they could recover before the people moved back.

In the Pacific Northwest, tribes closely followed the life cycle of the salmon and restricted their fishing to certain times and places. Catching and preserving salmon when the fish were in a particular phase of their life cycle meant that different foods could be prepared. The First Salmon was designed to honor the salmon, and, depending on the time of the season when certain species came back up the rivers to spawn, instructions would be given concerning the length and intensity of the tribal fishing activities.

Some tribes fired woods and grasslands from time to time to burn away the underbrush and ensure new and better grass for the next year's growth. Burning of grasslands was not arbitrary but depended upon the knowledge of how grazing animals had themselves used the lands before. People recognized that some animals did not like certain areas and tended to avoid them. It would have been useless to burn these areas with the hopes of enticing the animals to these lands.

In the Southwest, irrigation was practiced intensively—the Pimas, Maricopas, and Tohono O'odhams using the natural contours of the land wherever possible rather than irrigation ditches. They recognized that ditches would often fill up with debris and that they didn't mimic the natural flow of water. Old men would observe how the waters ran off after a rain, and they tried to create similar channels by removing obstacles to water flow so that rains would naturally water their fields. Some channels were blocked with stones to create a fanlike effect so a great amount of land could be irrigated.

Finally, Indians recognized the important functions of other creatures in a landscape and tried to minimize their exploitation of them. Beavers were understood as the critical animals in many areas because they built dams that retained water and made the surrounding areas very fertile. Until the fur trade, when the beaver were virtually exterminated, the species was generally left alone except when the need for food became critical.

On the whole, Indians tried to live within the parameters of bird and animal activity, recognizing that nature created a balance most of the time and that humans had to help preserve that balance. There were frequently prohibitions on the kinds of game that could be harvested. There are stories of how, even though the people were very hungry, the hunters allowed old buffalo or deer to lie undisturbed because they respected their age. It might be said that by giving priority to the needs of other creatures, humans were able to live prosperously in an environment because they respected what the birds and animals contributed to maintaining the productivity and fertility of the whole.

Given all this evidence, it is not difficult to conclude that wilderness is not so much a physical state but a concept of the mind. Societies from the Fertile Crescent to Europe developed a belief system that understood the natural world to be malevolent or at best an arena for the struggle to survive. They therefore treated lands and animals as things to be conquered and subdued. This attitude might have more to do with the ongoing political unification of the diverse societies that inhabited this region than it does with a divine revelation of ultimate reality, applicable around the world. Nonetheless, their belief system evolved into a religion that then marched from the Mediterranean to San Francisco Bay, bearing what it thought of as a torch of enlightenment. In its wake, trees, animals, rivers, and mountains became a wilderness to be feared, a wilderness to be controlled and tamed.

For American Indians—concerned neither with unifying different tribes nor spreading the good news of the Lord's work here on Earth—the very same landscapes that would soon be plundered by white men were allowed to become familiar, sustaining, and sacred. Because American Indians saw themselves as one of many species living on equal footing, they understood themselves as wrapped within the fabric of all life that constitutes our natural world—a fabric that cannot be harmed without harming ourselves.

The First Conservationists

Chris Madson

 The story of the American struggle to preserve wild spaces is often told through a canon of heroes—wilderness champions like John Muir, Theodore Roosevelt, Bob Marshall, Aldo Leopold, Olaus Murie, Howard Zahniser, and David Brower. Like many heroes, they're larger than life. We remember the power of their insight, the battles they won and lost, and the rich legacy they left behind, but all too often, we forget the men themselves. We know what they did but seldom take the time to find out why they did it. Lost in the modern recollection of the wilderness movement is a motivation that has driven many of its most effective advocates—hunting.

A close look at the history of wilderness politics shows a striking connection between wilderness and hunters. The philosophical foundation of modern wilderness preservation was laid almost entirely by hunters; the scientific and recreational justifications for wilderness have come largely from hunters; and the practical leadership at crucial points in the fight for wilderness has been provided largely by hunters.

The pattern is too consistent to be an accident. The connection between a love for hunting and a love for wilderness is older than the human species itself; it is a return to an outlook that governed our lives over uncounted millennia and ultimately shaped us as a species. It starts with that childish urge to hold something wild in our hands, but for true hunters, it soon becomes something far deeper. Ultimately, it is an understanding of the processes and relationships that are woven into the fabric of the land itself and how that land supports human life. This is not the postcard appreciation of the observer; it is the passionate insight of a participant.

The oldest records that we have of this fascination come from the images on the walls in places like Lascaux, Altamira, and Chauvet. Some of this cave art was created thirty thousand years[1] ago in a time when southern and central Europe supported an incredible collection of wild animals. Herds of mammoth, antique bison, rhinoceros, aurochs, ibex, horses, and reindeer were stalked by lions, brown bears, hyenas[2]—and men.

Physically, these were modern folk. The last of the Neanderthal line was gone by thirty thousand B.P.[3], and the human remains associated with cave-art cultures are hardly distinguishable from the bones we bury today. The cave painters were experts in the craft of shaping tools from flint and obsidian; they bent and sharpened bone and antler to make spear points that were tougher than the old stone points; they invented the spear thrower; they wore jewelry, and they buried their dead.

They also produced some of the most graceful murals in the history of human art. Most of these paintings, sculptures, and carvings show animals, and the details make it easy to identify the species being drawn and, in some cases, the sex of the animal and even its behavior. Some of the smallest carvings are no bigger than a quarter, while some of the paintings of big game are almost life-sized.[4]

Generations of specialists have struggled to extract a message from these works of art. It seems clear that they were much more than exercises in esthetics—some of the most spectacular work lies in cave passages hundreds of feet from any natural light. While the details of their meaning will probably always escape us, these are almost certainly the icons of a distant religion.[5]

Whether they are images of gods or just go-betweens to facilitate communication with gods, they capture the fundamental tension that has always existed between the human animal and the rest of creation. On one hand, the ancient hunters probably prayed to, or through, the beasts. On the other hand, the hunters desperately needed to kill these animals. Scattered through most of the caves are images of bears with spears driven into their flanks, bison with entrails spilling out of huge gashes, lions battered from the beatings they had taken from their creators.[6]

Stranger still are the figures of changelings, creatures with the heads of stags and the legs of men.[7] Are these shamans wearing costumes or do they represent some mythic hybrid between hunter and game? We'll never know for sure, but it seems likely that, with emerging consciousness,

Paleolithic hunters recognized an inescapable conflict in their lives—they were forced to kill what they worshipped. To the degree that they acknowledged their kinship with the rest of the natural world, the blood of their gods and their brothers was also on their hands. It was the original sin.

A LATER FALL

Humanity lived with that uncomfortable insight for a long time. If the customs of modern subsistence hunters are any guide to our past, we spent much of our time over the centuries apologizing to our quarry and the gods who provided or withheld fresh meat, judging our worthiness by the quality of our prayers.

Then, about twelve thousand years ago, we began to invent a new way of living and, in the process, a new way of interacting with the world; we could control it; we could ameliorate nature. Our control began with the sickle for harvesting the seeds of wild grasses and, later, the grain from domestic crops. In places like Abu Hureyra, Çayönü, and Çatal Hüyük in southwestern Asia, small groups of incipient farmers invented the plow, the clay pot, and eventually the city.[8]

The agrarian way of life allowed us to step back from the notion of complete dependence on the natural world. Our fields and herds were under our command, and, as long as there was enough rain, we could fool ourselves into believing that we had risen above the old hunter's ties to the earth. With this new-found self-confidence, we began to divorce ourselves from nature itself.[9] Our estrangement from wild nature, our fear of it, began with the first fence, not the first city.

And so the concept of "wilderness" was born. It was the unsettled land that began at the edge of our fields and had a will of its own. Farmers began to fear and hate this untamed place.[10] What had been home for a thousand generations of hunters became the enemy of a new breed of humans. With this change in attitude toward the wild, the ancient motivations for hunting split. Some of the old ways of thinking about the chase may have been lost altogether; others were fragmented and found advocates among different classes and vocations.

For the farming peasantry on the fringes of early civilizations in the Fertile Crescent, the hunt was a way to augment the meat supply and, at the same time, protect crops and flocks from depredation. There was a subsistence-like overtone to the exercise, but since the farmers didn't really depend on wild meat, they had no stake in the long-term health of wildlife populations and the habitats that supported them. They took as much as they could, and if, in the end, the wilderness was overrun and destroyed, they counted it as a victory and thanked a new pantheon of gods. This new concept of the relation between humans and the earth was reflected in the changing deities themselves—symbols of animals and their earth-bound power were steadily replaced by super-humans who spent much of their time in the heavens.[11]

A second kind of hunting developed in the budding civilizations of the Fertile Crescent. At Çatal Hüyük, an eight-thousand-year-old mural still decorates the entrance of a building archaeologists call a "hunting shrine."[12] The painting seems to show beaters driving deer, and there is a single red figure standing out from the rest. In a society of increasing specialization, this picture could easily be the earliest depiction of a "sport" hunter, a man of position taking game for his enjoyment or some ritual use.

This hunter of privilege became more and more prominent as the civilizations of the Middle East developed. Hunters on horseback or in chariots, well-armed and accompanied by dog handlers and drivers, pursued dangerous game for the challenge of it. In the eighth century B.C., Homer sang of the lord Odysseus: "[I]n the lead, behind the dogs, pointing his long-shadowing spear. Before them a great boar lay hidden in the undergrowth...and from his woody ambush with razor back bristling and raging eyes he trotted and stood at bay...Odysseus' second spear thrust went home...his bright spear passing through the shoulder joint; and the beast fell, moaning as life passed away."[13]

Two hundred years after Homer, Ashurbanipal, ruler of Assyria, left this message to all men who came after: "In my royal sport, I seized a lion of the plain by its tail, and at the command of Ninurta and Nergal, the gods in whom I trust, I smashed its skull with my own mace."[14] Here is the wilderness, not as home to Stone-Age hunters or a threat to small farmers, but as a worthy opponent testing the mettle of a leader.

These two kinds of hunting have stayed with us through the rest of civilized time. Over the centuries, the small landholder has continued to hunt for subsistence and to protect his property, probably hoping that the wilderness at his door would eventually go away. At the same time, men of privilege have often sought out a wilderness hunting experience as recreation and a rite of passage. They also tried to protect what they considered their wildlife from the masses by setting aside extensive tracts of land as inviolate hunting preserves for themselves. In the year 1225, for example, Henry III published his *Carta Forestae*, the Forest Charter, governing use of these reserves in England. It ended capital punishment for a peasant convicted of deer poaching but sentenced the poacher to imprisonment for "a Year and a Day" and banished him from the country if he could not pay his fine.[15]

On the far side of the world, Marco Polo found the Chinese ruler Kublai Khan managing a huge system of wildlife preserves. "At this place," he wrote on his return, "there is also a fine plain, where is found in great numbers, cranes, pheasants, partridges, and other birds. He derives the highest degree of amusement from sporting with gerfalcons and hawks, the game being here in vast abundance...Many keepers are stationed there for the preservation of the game..."[16]

This, then, is the conflict and contradiction we brought out of the Middle Ages: a tradition of "sport" hunting set against a tradition of hunting for meat, and the protection of property that could be imperfectly described as "utilitarian." A fear of wilderness threatening the safety and livelihood of small farmers set against an enjoyment of wilderness for recreation and challenge—a drive to tame the wilderness set against an interest in preserving it.

A NEW, WILD WORLD

These two themes—the fear of wilderness and the need to tame it—dominated not only the consciousness of immigrants to the New World, but also the many histories written about this tension and how it gave birth to the American conservation movement. William Bradford, one of the principals on the *Mayflower*, can be taken as an apposite example for all those anti-wilderness immigrants: "Besides, what could they see but a hideous and desolate wilderness, full of wild beasts and wild men—and what multitudes there might be of them they knew not...The whole country, full of woods and thickets, represented a wild and savage hue...What could sustain them but the Spirit of God and His grace?"[17]

Yet, even in that community of devout Christians, raised largely in

English cities, there were other views. Thomas Morton was a son of London gentry. Trained in law and the classics, he had shown a taste for poetry and the chase in England. Soon after he stepped off the *Mayflower*, he developed warm relations with the local tribes and spent much of his time hunting and socializing with them. In the end, his behavior so scandalized the Pilgrims that they arrested him, threw him in the stocks, and finally shipped him home.

Here is Morton's impression of Bradford's "hideous and desolate wilderness:" "The more I looked, the more I liked it. And when I had more seriously considered the beauty of the place, with all her fair endowments, I did not think that in all the known world it could be paralleled. For [there are] so many goodly groves of trees, dainty fine round rising hillocks, delicate fair large plains; sweet crystal fountains and clear-running streams that twine in fine meanders through the meads...Contained within the volume of the land, fowls in abundance, fish in multitude, and discovered besides, millions of turtledoves on the green boughs; which sat pecking of the full ripe pleasant grapes that were supported by the lusty trees; whose fruitful load did cause the arms to bend...'Twas Nature's Masterpiece: If this land be not rich, then is the whole world poor."[18]

It was Bradford's suspicion of the wilderness that dominated American thought over the next two centuries. But it was Morton's enthusiasm that survived and ultimately flourished, and I suspect there were more people of Morton's bent than history records. Many of them were illiterate—they spent their lives, like Daniel Boone, in the fastness of the new West, far from intellectual debates on man's place on the new continent, hunting, trapping, and finally dying beyond the reach of the civilization that had spawned them. And among these far travelers was a different kind of frontiersman who searched for a new understanding. Perhaps the earliest was Mark Catesby.

Born in Essex in 1682, Catesby was the youngest son of a middle-class Englishman. After studying with a free-thinking minister and an uncle who had a passion for natural history, he came to the colonies in 1712, looking for new plants and a little adventure. He found both. His travels took him across the territories of Virginia and the Carolinas where he collected native plants, birds, and observations on the natives, turning his experiences into six notable books on the natural history of the Southeast, which appeared from 1730 to 1743.

Catesby was also an enthusiastic hunter who apparently shot for sport as well as specimens. In the preface to *Natural History*, he spoke of field trips into the Carolina wilderness: "This encouraged me to take several Journeys with the Indians higher up the Rivers, toward the Mountains, which afforded not only a succession of new vegetable Appearances, but the most delightful Prospects imaginable, besides the Diversion of Hunting Buffaloes, Bears, Panthers, and other wild Beasts."[19]

Nearly all of America's eighteenth-century field biologists came from the same mold. The Quaker botanist, John Bartram, and his son, William, roamed the eastern wilderness from Florida to Canada in search of new blossoms, birds, and mammals, and they did much of their collecting with guns. After these pioneering sportsmen-naturalists came George Perkins Marsh, Thomas Nuttall, Thomas Say, John Townsend, David Douglas, and a score of others, all of them with an active interest in hunting and/or fishing. Trained observers with a love of the outdoors, these men clearly saw the ravages that were being visited on the continent.

One of the most influential of this group was the artist, anthropologist, and natural historian George Catlin, who in the 1830s traveled up the Missouri River as far as the Mandan villages and the mouth of the Yellowstone. In 1832, he called for a huge wilderness reserve on the upper Missouri to protect the natives and herds of bison that were already "rapidly wasting from the world."

The time had come, said Catlin, for "a magnificent park...a beautiful and thrilling specimen for America to preserve and hold up to the view of her refined citizens and the world, in future ages! A nation's Park, containing man and beast, in all the wild and freshness of their nature's beauty!"

In 1843, sixty-three-year-old John James Audubon, another naturalist who collected with a shotgun, followed in Catlin's footsteps on the upper Missouri and made a similar observation: "Even now there is a perceptible difference in the size of the herds, and before many years the Buffalo, like the Great Auk, will have disappeared. Surely this should not be permitted."[20]

These observations were the leading edge of a groundswell of concern for America's wild places, its spokesmen including some of the leading intellectual figures of the time—men like James Fenimore Cooper,

Washington Irving, and Henry David Thoreau. Although the romantic views of Jean Jacques Rousseau marked the wilderness philosophies of all these men, there was also a sportsman's first-hand perception of the loss of game, fish, and their haunts. Cooper, Irving, and Thoreau all took a sporting interest in the outdoors,[21] which led them, along with thousands of other Americans, to decry the wholesale destruction of wilderness.

AMERICA'S FIRST WILDLAND RESERVES

As the American juggernaut ground west, others came to Catlin's conclusion, that government should do something to preserve examples of the wild continent. In 1864, the state of California set an example for the nation when it claimed the Yosemite Valley for "public use, resort, and recreation for all time."[22] Surprisingly, a New York landscape architect, Frederick Law Olmsted, moved to California just in time to be one of the prime movers in the effort to protect Yosemite.[23] Not so surprising was the fact that Olmsted was a hunter.[24]

The protection of Yosemite was followed in 1872 by the creation of America's first national park—Yellowstone. Many men were instrumental in establishing and protecting Yellowstone—naturalist and explorer Ferdinand Hayden, Judge Cornelius Hedges, the writer and lecturer Nathaniel Pitt Langford, U.S. Senator George Vest, Congressman Samuel S. Cox, and zoologist and magazine editor George Bird Grinnell—all of them dedicated to the region's wild landscapes and big game, and all of them sportsmen.[25]

New York State followed the Yellowstone example by setting aside Adirondack State Park in 1855, with "Adirondack" Murray, Verplank Colvin, and William J. Stillman all playing major roles in the fifty-year effort to protect these northern mountains. It was clear again that their enthusiasm arose, at least in part, from a long interest in hunting and fishing. *The New York Times* spoke for many people when it described the proposed park as a "people's hunting ground."[26]

Then, in 1891, the nation's forestry chief, Bernard Fernow, convinced the secretary of the interior to intercede with President Benjamin Harrison to seek legislation that would give him the power to establish the country's first national forests. The bill slipped through Congress, and Harrison promptly set aside 13 million acres of woodlands, some of which are now designated wilderness. Fernow and Harrison were both hunters.[27]

CONSERVATION BECOMES MAINSTREAM

As the century drew to a close, the notion of setting aside wildlands for public use gained momentum. When William McKinley was assassinated the stage was set for one of the most significant breakthroughs in the conservation of wilderness—Vice-President Theodore Roosevelt, a keen outdoorsman, rancher, and hunter, entered the White House.

A sickly child, Roosevelt spent much of his time studying birds and hiking and snowshoeing to gain the strength he envied in others.[28] Here began his lifelong obsession with the outdoors. As soon as he was able, he also took up hunting with the intensity of a young man with something to prove, and it remained his passion until he died.

The contributions Roosevelt made to conservation during his administration are legendary. He established what is widely regarded as America's first national wildlife refuge, Pelican Island in Florida, and followed up with scores of others, effectively launching the national wildlife refuge system. He created several national monuments, including the Grand Canyon, which was being overrun by unscrupulous concessionaires. He also added 130 million acres to the national forest system, and with Gifford Pinchot, an avid angler and head of the new Forest Service, he expanded the agency's role, stressing the need for professional forest managers.[29]

At the same time, the Pinchot family funded a graduate program in forestry at Yale University in 1900.[30] One of its first graduates was a man who would also leave an indelible mark on American conservation—Aldo Leopold, who grew up along the Mississippi River in Iowa, learning his hunting and fishing skills and ethics from his father, Carl, a principled sportsman who was instrumental in local wildlife conservation efforts. But it wasn't until Leopold began his career as a forester in the Southwest that his vision of preserving America's wildlands began to take shape. He saw grizzly bears and wolves killed and wrote of watching "a fierce green fire dying" in the eyes of one female wolf. At that moment, he came to a realization that never left him throughout the rest of his life: "I thought that because fewer wolves meant more deer, that no wolves would mean hunters' paradise. But after seeing the green fire die, I sensed that neither the wolf nor the mountain agreed with such a view."[31] Leopold also saw roads infiltrate the forests he loved and development creep along the rim of the Grand Canyon, and he bemoaned the change with words as true now as they were at the beginning of the twentieth century: "Man always kills the thing he loves, and so we the pioneers have killed our wilderness. Some say we had to. Be that as it may, I am glad I shall never be young without wild country to be young in. Of what avail are forty freedoms without a blank spot on the map?"[32]

In 1919, Leopold met Arthur Carhart, a young landscape architect with the Forest Service who cherished a deep interest in hunting and the outdoors. Their friendship was to have far-reaching implications for wilderness preservation. Like Leopold, Carhart had experienced a conversion, spending the previous summer surveying cabin sites on Trapper's Lake in the Colorado high country. The Forest Service planned to lease the cabins to the public, and, by the end of the summer, Carhart came back with an entirely different and noncommercial proposal. A place like Trapper's Lake, he thought, should be off-limits to cabins, roads, and "development" of any kind. It should be a new kind of reservation, set aside for recreation in a completely untrammeled setting.

Carhart's view of these pristine lands closely paralleled Leopold's thoughts on wilderness conservation, and Leopold asked the younger man to commit his ideas to paper. Carhart complied, and his "Memorandum for Mr. Leopold, District 3," dated December 10, 1919, became the earliest written description of the modern wilderness concept.

"These areas can never be restored to the original condition after man has invaded them," Carhart wrote, "and the great value lying as it does in the natural scenic beauty should be available, not for the small group, but for the greatest population. Time will come when these scenic spots, where nature has been allowed to remain unmarred, will be some of the most highly prized scenic features of the country."[33] Carhart convinced his superiors to exclude development from the area around Trapper's Lake, and, in 1922, he prevailed on supervisors of the Superior National Forest in northwestern Minnesota to extend the same protection to the lake region that later became the Boundary Waters Canoe Area.[34]

In the meantime, Leopold was working to establish a wilderness preserve in the forest holdings of New Mexico. He had his eye on the headwaters of the Gila River, a rugged, remote piece of high country in the

Gila National Forest. In his 1924 plan for recreation in the forest, he proposed a 775,000-acre reserve in which there would be no roads or other development.[35] In June of that year, the Gila became the first formally recognized wilderness area in the United States, and Leopold wrote: "Why should not the Gila, and other similar areas, if possible one in each western state, be declared permanently roadless, and dedicated to that particular form of public recreation beloved by the wilderness hunter?"[36]

When the young Robert Marshall, a trained forester and explorer of Alaska's Brooks Range, decided to found an organization dedicated to wilderness preservation, it seemed natural that he would turn to Leopold for assistance. Leopold gladly consented to joining the infant Wilderness Society but was too busy with his game management studies in Wisconsin to take on the presidency of the organization.[37] However, in the first issue of the organization's magazine, he penned telling words. The Wilderness Society, he wrote, implied "a disclaimer of the biotic arrogance of homo americanus. It is one of the focal points of a new attitude—an intelligent humility toward man's place in nature."[38]

Leopold kept refashioning this idea, saying only a year before his death in 1948, "Ability to see the cultural value of wilderness boils down, in the last analysis, to a question of intellectual humility. The shallow-minded modern who has lost his rootage in the land assumes that he has already discovered what is important; it is such who prate of empires, political or economic, that will last a thousand years. It is only the scholar who appreciates that all history consists of successive excursions from a single starting-point, to which man returns again and again to organize yet another search for a durable scale of values. It is only the scholar who understands why the raw wilderness gives definition and meaning to the human enterprise."[39]

Following Leopold's death, the movement to preserve wilderness passed into the hands of many others. Biologist and hunter Olaus Murie, a good friend of Leopold's, served as director of The Wilderness Society from 1945 until 1962, dying just before the passage of the Wilderness Act, a federal law he had spent most of his career advocating.[40]

In the ensuing decades, other hunters continued to put their stamp on wilderness preservation: Sig Olson, the champion of the fight to save Minnesota's Boundary Waters, came to the wilderness as a boy with an insatiable interest in hunting and fishing; Stewart Udall,[41] interior secretary in the Kennedy and Johnson administrations and a lifelong advocate of wildness on the American landscape, also began his love affair with the outdoors hunting and hiking the hills around his home in Arizona; and President and hunter Jimmy Carter added 56 million acres in Alaska to the wilderness system during his term in office.[42]

These are the names we remember. Many others will never occupy more than a footnote in the books—men like Cecil Garland, the Montana hunter who after his first September in the high wilderness swore that "whatever the cost for whatever the reason, I would do all that I could to keep this country as wild as I had found it."[43] He was instrumental in the creation of Montana's Scapegoat Wilderness in the early 1970s.

Finis Mitchell and Paul Petzoldt left their marks farther south. Mitchell, a hunter in his early days and an enthusiastic angler all his life, stocked the lakes of the Wind River Range and introduced three generations of hikers to one of Wyoming's finest wild places. Mitchell ended his guidebook, *Wind River Trails*, with a promise: "I shall endeavor with all my ability and steadfast efforts to preserve and add to our wilderness."[44] Petzoldt, an enthusiastic hunter and angler, translated his passion for climbing and wilderness into a commitment to education. He established the National Outdoor Leadership School, an institution that has trained thousands of back country users and instilled an appreciation of wilderness values.[45]

MODERN HUNTERS—ALLIES AND OPPONENTS OF WILDERNESS

There is an altruistic thread running through the sportsman's support for wilderness that Aldo Leopold crystallized in his land ethic. "A thing is right," he observed, "when it tends to preserve the integrity, stability, and beauty of the biotic community. It is wrong when it tends otherwise."[46] This standard implies a need for technical understanding, careful study, and a highly developed aesthetic sense. It is a high-minded goal, one that Leopold himself suspected we would struggle to reach. "We shall never achieve harmony with the land, any more than we shall achieve justice or liberty for people," he concluded. "In these higher aspirations the important thing is not to achieve, but to strive."[47]

These "higher aspirations" are part of every facet of the conservation movement, including the efforts hunter-conservationists have made and continue to make on behalf of nongame species, wildlife habitat, and wilderness. However, there is another side to hunting—what I've called the "utilitarian" view. Hunters of this ilk care very little about process; their goal is to get some shooting, possibly some meat, occasionally a set of large antlers as quickly as they can with as little effort as possible. They're not much concerned with the setting of their hunt, whether they're crowded or alone, whether the animal they're hunting has spent its entire life in the wild or has been raised behind a fence.

Over the last century, the utilitarian hunter has had little patience with the concept of wilderness. He sees roadless areas as inefficient, exclusionary, and ultimately elitist. He generally regards any conservation effort on behalf of nongame as inefficient and recent efforts to reestablish large predators in the West as wasteful; in fact, he is likely to be a staunch advocate of large-scale predator control.

And so hunters remain divided, as we have been since the advent of agriculture. There is still reverence for wildlife and wild places among hunters, and the sense of gratitude and obligation. There is also arrogance, the desire to possess, and the temptation to kill for the sake of killing.

The great divide in the hunting community is nowhere more apparent than in the response to the Forest Service's Roadless Area Conservation Rule. Proposed by the Clinton administration, this rule bans the construction of new roads into roadless areas on national forests. Bitter voices were raised against this effort to head off new road construction, and some of the proposal's most vocal opponents were hunters. A spokesman for Safari Club International, an organization devoted to trophy hunting, testifying before a House subcommittee, said, "We as sportsmen question the intent of a moratorium"[48] on road construction in national forests. The National Rifle Association put it more simply: "No roads; no access; no hunting."[49]

In this debate, as in so many others over the last thirty years, many people have claimed to speak for all hunters. A salient example is Craig Thomas, Wyoming's Senator, who attacked the Clinton roadless plan because it "would ultimately restrict access by recreationists, hunters, fishermen, and other responsible multiple-use enthusiasts."[50] All too often, mainstream news coverage of the debate has also taken up such assertions without bothering to confirm them. *Time Canada* offered a typical example in an article on the controversy surrounding fires and roadless areas in the

American West. "The argument over roadless lands is thus framed in vastly different perspectives of time," the authors pointed out. "Opponents— off-road-vehicle enthusiasts, hunters, fishermen, loggers, miners, and organized labor (citing the loss of jobs)—do not speculate centuries in advance. They want use of the land now."[51] *Time Canada* is half right— hunters do want use of the land, but they also care about how it is used.

To that end, devotees of Roosevelt and Leopold's legacies are organizing. Last year, the Theodore Roosevelt Conservation Alliance (TRCA) was created by a group of founding trustees—the Izaak Walton League of America, the Mule Deer Foundation, Rocky Mountain Elk Foundation, Trout Unlimited, Wildlife Forever, and the Wildlife Management Institute. Representing a coalition of sportsmen who want sound management of public lands, TRCA commissioned a survey to find out how these sportsmen felt about roadless areas. Eighty-six percent of anglers and 83 percent of hunters "supported efforts to keep the remaining roadless areas in national forests free of roads."

TRCA director Bob Munson had this to say about the statistics: "The fact that hunters and anglers desire to protect wild country and have access to these areas is not contradictory. Experiences in such areas are the gold standard by which the best hunting and fishing experiences are judged. The roads debate provides the opportunity for hunters and anglers to champion the value of remote areas..."[52]

Evidence from other parts of the country indicate that Munson also speaks for a vast majority of today's hunters. One researcher who has corroborated his views is Dr. Chris Potholm, professor of political science at Bowdoin College, who in the early 1980s assessed public opinion for The Nature Conservancy in Maine. He discovered a heartening overlap in the views of hunters and members of the environmental community. When organizations representing hunters and those representing environmentalists recognized their common ground, Potholm found "an unassailable coalition" of up to 70 percent of voters. His subsequent experience in Nevada, Arizona, and New Mexico show a similar margin of support for the environmental agenda, including wilderness.[53]

Such a coalition of hunters and environmentalists could have enormous implications for the future of wilderness politics in the upcoming decades. As Tim Richardson, TRCA's deputy director puts it, "Since sportsmen are in the political center, you can then achieve a conservation objective no matter which party controls Congress and the White House. What you get is a bipartisan conservation majority that can survive the partisan swings that we're in right now."[54]

THE FUTURE OF HUNTING

The debates over wilderness and the blood sports will intensify in the coming decades. As the U.S. population edges toward 400 million, we will find that anachronisms like wilderness and hunting cost more to keep and are more controversial as a result. One side in this debate has little patience with what they see as a preoccupation with the past. They believe that change is synonymous with progress, that we should get on with the business of imposing human order on the unruly people and places that have resisted domestication over the millennia.

These apologists for the cult of "development" have a powerful voice in modern culture, but they don't speak for the majority of us. At some instinctive level, most of us still feel a need for wildness. The evidence is subtle but pervasive. How else can we explain our attachment to our pets, our inclination to decorate our homes with flowers? Why do we build fountains in our public places and set aside parks in the middle of our largest cities? What draws our eyes to the heart of a fire, and why are we still afraid of the dark? We have not left the Pleistocene behind. The mark of the wilderness is on us still.

The Canadian biologist C. H. D. Clarke once wrote: "In the flight from nature, the hunter…is being dragged along unwillingly…The more hunters there are, the slower that flight will be. Call us a drag if you like, but a good sea anchor keeps your head to the winds of change. Hunting is living. Living is an art. In our much-vaunted progress, we confuse technology with the art of living."[55]

Like Clarke, those of us with an affection for wilderness know there is a fundamental difference between standard of living and quality of life. Hunters may know it best of all. Quality hunting—requiring a fragile combination of abundant wildlife, untamed land, and solitude—is one of the most sensitive indicators of environmental health. When that wild experience is eroded, it affects the hunter on a spiritual and practical level; the fate of hunting and the fate of the wilderness cannot be separated.

This is why the best of hunters will always be a force for wilderness, and it is why wilderness advocates should look to the 17 million members of the North American hunting community for support. Hunters and non-hunters alike, we are the keepers of a tradition as old as the paintings at Chauvet and Lascaux. Together, we can remind a nation that has lost its memory of the connection between people and wilderness. We can never be truly happy without wild places—they gave us life in the beginning, and they nurture us still.

PART II

THE HUMAN LANDSCAPE

It may take time for any culture
to become truly "native," if that term is
to imply any sensitivity to the ecological
constraints of its home ground.

— GARY PAUL NABHAN

"Gifts of Nature" in an Economic World

Thomas Michael Power

∽ Commercial development of America's last remnants of unprotected wilderness amounts to the destruction of that which is increasingly scarce and can never be replaced for that which is common and easily reproduced. In the most hard-nosed economic sense, this makes no sense.

Yet in the daily tussle of politics, we regularly hear that "saving" wilderness will "lock up" natural resources and lead to massive job losses and economic decline in dependent local economies. I put "saving" in quotes, for one fact too often lost in this discussion is that no local economy presently depends upon these resources since they have never been commercially developed. These wild places have remained roadless and unexploited because of their remoteness, ruggedness, and sparse commercial values. They are unique, non-reproducible "gifts of nature," increasingly scarce and therefore of increasing value for a variety of qualities, only one of which is wilderness appreciation. Pure and simple, they also have economic value as wildlands.

The opponents of wilderness preservation get the arguments about the economic value of wilderness backward. Their reasoning goes like this: Wildland preservation puts valuable natural resources—timber, minerals, energy resources and motorized and developed recreational opportunities—out of reach of commercial use, burdening the economy, sacrificing jobs and earnings, and leaving the community poorer than need be. If we want jobs and a growing local economy, we thus should press on to develop the last of America's still-pristine but unprotected wildlands.

Not only is this argument usually wrong, the reverse is actually often the case: Wildlands left undeveloped have a higher economic value than those that are exploited for commodities. This is one subject of my essay.

The other is that an accurate economic analysis of the value of wilderness areas can be a powerful tool for their preservation.

WILDLANDS AND THE LOCAL QUALITY OF LIFE

Let us begin by reviewing the hard data. Economic research has shown that areas with intact natural environments, protected by official wilderness or park status, have attracted higher levels of economic activity than otherwise comparable areas without intact natural environments. Here is a brief summary of the evidence:

- Rural, non-metropolitan counties in the Western states with more than 10 percent of their land in national parks, monuments or wilderness saw job growth 1.85 times the average for Western non-metropolitan counties; income grew 1.43 times faster.[1]

- Unprotected wildlands yet to face roaded development also appear to attract economic activity. In one study, the acreage of U.S. Forest Service inventoried roadless areas within fifty miles of a county's center was positively correlated with employment and income growth. The strength of that correlation increased as the analysis shifted from all counties to just the non-metropolitan counties (no cities larger than 50,000) and then to the purely rural counties (no cities greater than 2,500).[2]

- Another researcher found strong correlations between protected federal wildlands as a percentage of total county land area and employment, per capita income, total aggregate income and population growth.[3]

- Researchers puzzled by the growth of population in western Montana despite low wages and incomes studied the location of new residential housing to explain decisions homebuilders were making. The closer a location was to a designated wilderness area, it turned out, the higher the likelihood of new construction. The same was true of national parks. Distance to Montana's larger population centers and access to major highways were also important.[4]

- The existence of designated federal wilderness enhanced nearby land values, according to an analysis of the value of over six thousand land parcels that were transferred in Vermont's Green Mountains. Parcels of land in towns near designated wilderness sold at prices 13 percent higher than in towns not located near wilderness.[5]

- A recent University of Maine analysis of migration patterns in the Northern Forest region (Minnesota, Wisconsin, Michigan, New York, Vermont, New Hampshire, and Maine) found that, in general, jobs were following people's residential location decisions rather than people simply moving to where employment opportunities were. And in-migration was greater in those counties with a higher percentage of their land base set aside for conservation purposes such as state and national forests, parks, and wildlife refuges. As a result of the higher in-migration, job growth was also higher.[6] Given that timber harvests were falling on federal conservation lands during this time period, the positive impact the presence of these lands had on in-migration and employment was impressive.

- No negative impact of wilderness designation on employment or income has been found in analysis of the impact of federal wilderness areas and national parks in the Mountain West.[7]

The conclusion: In areas with only small cities and towns, the more of the land base that was in national wilderness, parks and monuments, the higher were the measures of local economic vitality. Protected wildlands drew new residents who were willing to sacrifice a certain amount of income in order to live amid the higher quality natural environments that they perceived federally protected landscapes provided.[8]

Other research has shown that protecting natural landscapes also positively affects the location decisions made by business firms. With the shift from goods production to knowledge-based services (such as those involved in research, insurance, finance, and high technology), more firms have become relatively "footloose." The success of these companies is less dependent on location than on obtaining and keeping the highly qualified personnel they need at a reasonable cost. Protected wildlands appear to draw this kind of generally "clean" economic activity to nearby communities.[9] As a result, natural amenities have become an important part of a region's economic base. As one recent study of the role of environmental quality on the location of high tech firms put it:

Knowledge workers—scientists, engineers, software developers, computer and communication technicians—essentially balance economic opportunity and lifestyle in selecting a place to live and work. Thus, lifestyle factors are as important as traditional economic factors such as jobs and career opportunity in attracting knowledge workers in high technology fields…The new economy dramatically transforms the role of the environment and natural amenities from a source of raw material and a sink for waste disposal to a key component of the total package required to attract talent and in doing so generate economic growth.[10]

This amenity-supported economic vitality is not an unalloyed blessing for our wildlands, it is important to note, since population growth mushrooms in adjacent rural areas, often with little or ineffectual local land use planning. The point is not to praise this pattern but to recognize it is as a national phenomenon that is shifting the base of local economies away from natural resource exploitation. We cannot wish this trend away, but must stress the need to protect the wider local environment, even as we focus on preserving the wildlands that are a key to attracting this new rural population.

COUNTERING THE ECONOMIC CLAIMS OF WILDERNESS OPPONENTS

In combating wilderness opponents' exaggerations of negative local economic impacts, it is important to counter with an effective critique of their claims. Here are some key steps in that critique.

Don't Define "Local" Too Narrowly. One way opponents project very large negative impacts of wilderness designation is to focus very narrowly on small cities or rural areas. The small size of the "local" area makes the loss of almost any jobs, such as those in logging and other forest products, appear relatively large, serving the political purposes of wilderness opponents.

In general, it does not make economic sense to analyze a small community as if it were a stand-alone economy. Those who live in a small town often do not actually work there; those who work there may not live in that town; and those who shop there may neither work nor live in the town. The only way to take into account all of that mobility, and keep track of where individuals work, shop, and live, is to define a relatively large functional economic area that contains within it most of that mobility and commuting and, as a result, most of the economic links. In this larger interconnected economic area, the impact of changes in local natural resource employment is likely to be relatively much smaller.

Place Estimated Impacts in an Appropriate Quantitative Context. The bald assertions made by wilderness opponents must be put into an appropriate quantitative context. If, for instance, it is estimated that preserving a wilderness area will have a net impact of thirty jobs being lost and a reduction in local payroll of $750,000 per year, one obvious analytical question is whether this is, in fact, a large local impact. The loss of thirty jobs in an area that has total employment of only three hundred would represent a much greater negative impact than the loss of thirty jobs in a local economy employing one hundred thousand which has been adding three thousand jobs a year or about thirty jobs every four days. In this latter context, the loss of thirty jobs might be considered easily "digestible," the kind of dynamic change that occurs in the private economy every few days. Comparing the estimated local economic impacts to both the overall level of employment and income and the ongoing annual change in them helps establish a useful context for interpreting the economic impact data.

This can be done by expressing the change in terms of total employment or total income and also by expressing it in terms of the number of days of normal economic growth that the change represents.

Place the Estimated Impacts in the Context of the Changing Economy. Ongoing change in the economy means that the relative importance of various types of economic activity is constantly changing. Some industries decline in importance while others grow in importance. Industries that are the source of local economic vitality also shift. The sources of new jobs, additional income, and incremental government tax revenues now and in the future may be—and almost certainly are—quite different from the sources of local economic vitality twenty or thirty years ago.

The local conventional wisdom about the economic base is often a rearview mirror perspective, a romanticized, inaccurate but widely shared understanding tied to historical experience rather than contemporary reality. Local opinion-makers, elected officials, and other "influentials" may all share this view.

Wilderness advocates should use the factual information available on the *actual* local economy to help change this erroneous but widely shared conventional wisdom about the keys to the local economy's future. When the ongoing changes in the local economy are considered, it may become clear that future sources of employment and income lie in altogether different economic sectors than the natural resources sectors that depend on the roading and developing of local wildlands.

One method to quickly determine how the local economy has been changing over time and what the sources of local economic vitality have been is to use simple trend analysis of the economy and its chief components. One can look, for instance, at how employment levels by major industrial category have changed over the last twenty-five years.[11] One can also compare the trend line for the industries that are assumed to be the economic base for the local area with the trend line for the rest of the economy that the assumed economic base is supposed to be driving.

Figure 1
Federal Timber Harvest and Wood Products Employment
Compared to Jobs Outside the Wood Products Industry: Montana

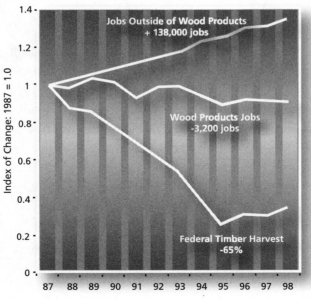

If the assumed economic base and the rest of the economy are moving in different directions or in different patterns, that is a warning that the assumed local economic base may be seriously misdefined. Figure 1 sketches this out for Montana, its wood products industry, and federal timber harvests. As federal timber harvest declined dramatically and employment in wood products declined modestly, employment throughout the rest of the economy dramatically expanded rather than following wood products downward.

Analyze the Whole of the Local Economy, Not Just Part of It. It is often tempting to use informal understanding of the local economy to reduce the complexity of the economic analysis that is required. When wildland protection opponents do this, they skew the argument to their favor, and wilderness advocates should call them on it.

If, for instance, it is assumed that the local economy is primarily a forest products or mining or agricultural economy, the analysis may be limited to just studying impacts on that one sector. This kind of shortcut is inappropriate. All sources of employment and income in the local economy should be considered.

Figure 2
Sources of Income in the Montana Economy, 1998

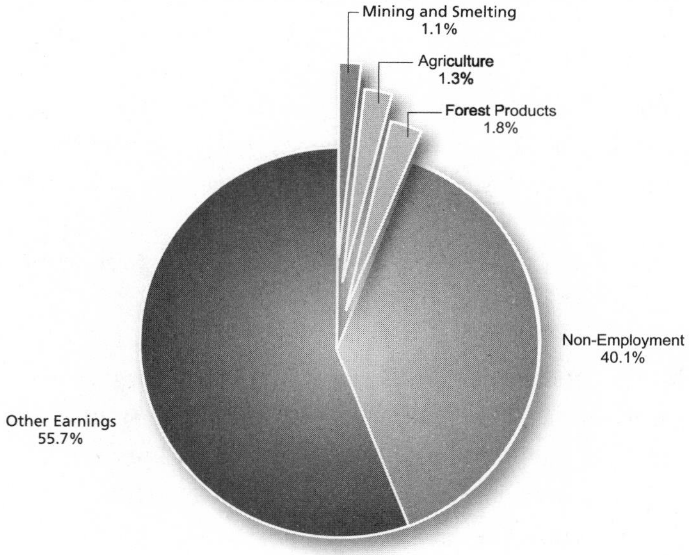

Mining and Smelting
1.1%

Agriculture
1.3%

Forest Products
1.8%

Non-Employment
40.1%

Other Earnings
55.7%

Often local economic analysis focuses only on wage and salary income. But wage and salary income may represent less than half of local income. Wage and salary income ignores self-employment income and non-employment income (social security, pensions and investment income such as dividends, rent, and interest). Ignoring half of local income coming from sources driven by forces quite different from those driving wage and salary income is a serious economic error, putting wildland protection advocates on the defensive unnecessarily. Figure 2 demonstrates how dramatic a story the broader data can tell. It shows the sources of income in the Montana economy, an economy usually described as "natural resource dependent."

Application of the four steps outlined here so far can dramatically change the interpretation of local economic impacts. Impacts that commercial development interests claim will be huge can actually be shown to be trivially small.

Consider the impact of the January 2001 Clinton Roadless Area Conservation Rule in Montana. Because almost 6 million acres of Montana National Forest roadless areas are set off limits to timber harvest under the Roadless Area Rule, timber interests argue that the rule will devastate the state economy and "timber dependent" small communities.

Gathering a few basic facts allowed Montana wildland advocates to show that this was not the case. Those facts included the following data for 2000:

- Percentage of total Montana jobs associated with the wood products industry in those counties adjacent to roadless areas: 3.4.
- Percentage of total Montana timber harvest from National Forest land: 15.
- Percentage of National Forest timber harvest in Montana coming from or planned to come from roadless areas: 2.

Multiplying these three small percentages together yields a truly small percentage that represents the overall jobs impact of setting almost 6 million acres of National Forest off limits to logging: 3.4 percent x 15 percent x 2 percent = 0.01 percent or one-hundredth of one percent. (See the sidebar on page 71 for a complete discussion.)

There are several steps in addition to the four we have listed so far that need to be incorporated into any complete economic analysis of wildland preservation:

Include Impacts on Other Local Businesses. It is important not to focus on just one particular industry that will be negatively affected and ignore the impacts that are felt elsewhere in the economy because of the environmental connections. Protection of wilderness will be felt by other businesses (outfitters, businesses servicing outdoor recreation and visitors, irrigators and municipal water users, commercial fishing, etc). The negative impacts on extractive industries may be partially, fully, or more than offset by positive impacts on other businesses, and vice versa.

Include "Indirect" Impacts of Changes in Local Environmental Quality. People regularly make sacrifices in pursuit of higher quality living environments and business leaders know that attractive areas can draw the needed work force with less upward pressure on wages. The local quality of life is thus part of any community's economic base—and this includes environmental quality and nearby wildlands.

Improvements in local environmental quality understandably tend to stimulate economic activity while degradation of local environmental quality tends to retard it. This "indirect" impact of changes in environmental quality on local business activity needs to be included in the overall evaluation of local economic impacts of wildland preservation.

Be Cautious of "Multipliers." Changes in local expenditures will cause "ripple" effects among other local businesses. These "secondary" impacts are often summarized through the use of a multiplier: For instance, the gain of a million dollars in new payroll or thirty new jobs (the direct impacts), is said to lead to $3 million in total new income or 120 total new jobs if the income multiplier is asserted to be 3 and the employment multiplier 4. "Multiplier effects" have been regularly exaggerated in the past for political purposes. There is almost never any follow up to check the accuracy of the "forecasted" job gains or losses. In fact both job "gains" and "losses" are almost always hypothetical jobs tied to a self-interested party's modeling of an economy frozen in time. The effect has been to exaggerate local economic impacts.

Put Estimated Local Impacts into the Context of an Entrepreneurial Economy. Local economic impact analysis often is completely static in character. It assumes that some "shock" hits the local economy—a wild landscape is "closed" to logging or oil and gas leasing—and purported "job losses" are assumed to be permanent, with population adjusting downward through out-migration due to the reduced set of economic opportunities. Taken to the extreme, this is the Oklahoma-in-the-Dust-Bowl imagery.

But market economies are known for their adaptability to changed conditions. Entrepreneurial response to change seeks to minimize the negative impacts and maximize the positive. As a result, valuable resources shift from one type of economic activity to another, generating economic activity in new fields as activity declines in previous economic pursuits. Labor, land, and capital rarely remain unemployed for long periods of time, one of the much-vaunted benefits of our free-market economic system.[12] As a result there are few permanent losses caused by economic change. It is the net change after these economic adjustments have been made that represents the relevant economic impact.

The adjustment of economies in the Pacific Northwest to dramatic declines in federal timber harvests during the 1990s provides a good example of the powerful self-adjusting nature of market-oriented responses in offsetting the impact of the reduced federal harvest.

In the state of Washington, federal timber harvests declined by 92 percent between 1988 and 1998. Total Washington timber harvests from all ownerships fell by 45 percent. Forest products production, however, remained stable while employment declined by only 14 percent.[13] The primary reason that forest industry production and employment were much more stable than total harvests was that the export of unprocessed logs from Washington to Japan—a practice long fought by environmentalists—declined by 70 percent and those logs moved through Washington mills instead of directly overseas.

Also, as harvest on federal lands declined, harvests from private, non-industrial timber lands grew, partially offsetting the federal decline. Finally, logs harvested in British Columbia were attracted by the higher prices being offered in Washington. These changes in log flows allowed lumber, veneer, and pulp production to remain stable in Washington despite the fact that timber harvests were nearly cut in half.[14] At the same time, job growth across the state generated more than 800,000 net new jobs, in no small part due to the attractiveness of the state's increasingly protected forested landscapes.[15] The ongoing economic growth in the Cascade region of Washington and Oregon dwarfed job losses in forest products, and high-tech employment now exceeds forest products employment. Interestingly, this growth took place in almost exactly the same areas where federal timber harvests were plummeting.

The important point is that markets respond to impending shortages in supply in ways that tend to reduce, if not eliminate, those shortages. Given the modest impact that the protection of most wildlands would have on total wood fiber supply, one can expect market adjustments in the flow of raw materials to relieve much of the projected impact of reductions in wood supply that might be estimated assuming a static, non-adjusting economy.

EVALUATING JOB "OFFERS THAT CANNOT BE REFUSED"

Inflated "job promises" are a staple weapon of commercial interests and chamber of commerce boosters in debates over wildland protection. Because natural resource jobs (especially in mining) can be among the highest paid jobs available, a commercial development proposal can plausibly be presented to a community as a solution to the low incomes, high poverty rates, and limited employment opportunities that are typically said to characterize rural areas. But the surface plausibility of this "conventional wisdom" masks a broad array of uncertainties and problems.[16]

In the public debate over wildland protection, job promises associated with proposed commercial development are almost always inflated and should be corrected for all of the following:

- "Possible" jobs are not the same as actual jobs. Many promised jobs associated with speculative projections may well not pan out.
- Many of the new jobs will go not to existing local residents but to in-migrating newcomers or commuters.
- Short duration jobs are not the same as permanent jobs. Development and construction period jobs are often much more numerous than the

long-term production jobs.

- The stability of employment is important. Natural resource jobs are often at the mercy of national and international commodity market fluctuations.
- "Multiplier" impacts, as previously discussed, should be viewed with suspicion since they are subject to arbitrary manipulation and somebody's hypothetical modeling of the future.

The economic proposition that is presented to the local community in support of commercial development of surrounding wildlands usually contains the following seemingly powerful points, central to developers' propaganda:

- Pay and income in small cities and rural areas is almost always well below national and state averages. If pay is low, the area clearly lacks sufficient high paid jobs. Natural resource industries offer such jobs.
- In many non-agricultural rural areas and small cities, unemployment rates are above national and state averages and often have been so for many years. Creating more jobs locally will put the unemployed to work and reduce the unemployment rate.
- The negative version is that environmental restrictions on extractive industry's access to wildlands are a "lock up" of natural resources causing ongoing losses of high paid jobs, impoverishing communities and driving pay and income downward.
- Efforts to replace the natural resource jobs that have been lost with "new economy" jobs are bound to be unsuccessful. There really is no replacement for natural resource jobs. The "new economy" generates "service" jobs that are low paid and part-time, and tourism is seasonal.

As convincing as these economic arguments for the commercial development of wildlands may appear to be, they often are built on false assumptions and exaggerated claims. The larger historical reality is that past natural resource development has rarely converted areas that have embraced them into prosperous communities.

It is a startling fact that even though forest product and mining firms pay wages that are well above average, the towns whose economies are built around these industries are usually somewhat rundown, aging communities that show very little economic vitality. The natural resource extraction industries rarely bring prosperity to these local communities for several reasons.

First, these industries sell their products into national and global markets and face increasing competition from an increasingly globalized economy. This dynamic causes the supply and demand—and price—of the commodities produced to fluctuate widely. As this happens these firms regularly have to lay off their workers, interrupting the flow of income through the community. Pay is high but both pay and employment are increasingly unstable—a classic boom-and-bust economy.

Moreover, these industries have had a long time period for laborsaving technology to be brought to bear, raising labor productivities faster than in almost any other industry. Fewer and fewer workers are needed for any given level of extraction and production, so employment has steadily declined.

These extractive industries often operate in a non-sustainable manner. Timber harvests often exceed the growth of new commercial supplies, leading the industry to ultimately reduce their operations and move to other regions in pursuit of adequate supply. Mining companies have developed techniques that allow them to completely exhaust a mineral deposit in a matter of little more than a decade, and the same is true for oil and gas drilling. That means that the economic activity is temporary and does not support long-term local investment by local residents and businesses.

Finally, these industries often have a devastating impact on the local environment. The landscape may be permanently scarred, water and air are polluted, wildlife habitat and fisheries degraded, and wild places lost. This makes these communities poor draws for potential new residents and businesses and for tourism—yet another reason why the high wages paid in these extractive industries do not assure sustained community prosperity.

ARE SMALL CITIES AND RURAL AREAS
DESPERATE FOR WILDLAND DEVELOPMENT?
Low Pay and Income in Rural Areas. Although it may seem counterintuitive, low pay and income in a local area is not necessarily a sign of

poor economic well-being. Highest pay is found in the largest of our metropolitan areas; the lowest pay is found in our smaller cities and rural areas. Yet the population, voting with its feet, often moves out of high wage, densely settled areas and moves into lower wage small cities and rural areas. Many people are willing to accept lower pay in return for the lower cost of living and higher quality of life that effectively provides them with a "second paycheck," so low pay does *not* actually represent reduced economic well-being. Of course there are also low pay areas that are losing population, such as the Great Plains. But even there, the bright spots are high amenity areas such as the Black Hills.

High Unemployment Rates and Natural Resource Industries. It is ironic that natural resource industries use the high unemployment in their own communities as proof that additional natural resource development is necessary. High pay industries located in relatively low wage rural areas tend to create unusually high unemployment rates. The reason is that the high wage firms tend to draw many workers and their families hoping to obtain one of the high wage jobs. A hopeful but surplus labor force builds up in the area. In addition, instability in commodity markets and ongoing displacement of workers by new technology regularly leads to existing natural resource workers being laid off. The reservoir of workers hoping for future high paid employment in the natural resource industry continues and accumulates, raising the local unemployment rate. High unemployment is thus a characteristic of these industries. Increasing the community's dependence on such industries, then, will not solve the high unemployment problem; it is likely to make it worse.

The Local Mix of Industries Does *Not* Determine Local Pay Levels. During the 1980s, employment in natural resource industries declined as the nation suffered through back-to-back recessions that hit non-metropolitan areas particularly hard. As natural resource industries recovered, they often installed cost-reducing technologies that permanently reduced their need for workers. During this same period, average pay declined dramatically in non-metropolitan areas and did not begin to recover until the second half of the 1990s.

These two trends are regularly linked together as if one caused the other; that is, it is assumed that average real pay fell *because* jobs were lost in the natural resource sectors. But empirical analysis shows that this was not the case. Average pay fell because pay in almost all sectors of the economy declined, including, especially, pay in the natural resource sectors. It was downward pressure on wages across the board that caused the pay losses. That downward pressure came not from local sources but from national and international forces. If we were magically able to freeze the structure of jobs where it had been in some past "golden age," say, in the late 1960s or late 1970s, almost all of the pay loss would have been experienced anyway.

The Shift from Goods to Services Does Not Necessarily Harm Economic Well-Being. Despite the declines in employment in the natural resource sectors, overall employment in most local areas has been expanding significantly. Spokespersons for the natural resource sectors tell us these new jobs are inferior service jobs that pay poorly, offer only part-time work, and make little use of the skills and abilities of adult workers. They are "burger flipping" jobs that might be appropriate for kids but are inappropriate for adults trying to support families. The proliferation of these jobs, we are told, is what is impoverishing our communities. Industry boosters claim that the "new economy" is inferior to the old natural resource economy that provides full-time, high wage, long-term jobs.

This critique is built around categorical confusion. "Burger flipping" is not a service job; it is a retail trade job. "Services" include well paid professional activities such as medical, repair, engineering, education, computer, business, and legal jobs. Services also include some low paid jobs such as day care, hotel room cleaning, and domestic help. Service jobs have been growing faster than the overall economy; retail trade jobs have grown proportionately with the economy. There is a shift in jobs toward services but not toward "burger flipping" and other retail trade jobs. Although there are low paid service jobs, there are also many low paid manufacturing and agricultural jobs.

The shift from goods production to services has been underway since the earliest days of this country; the percentage of the population engaged in goods production has been steadily declining. There is little evidence that this shift has left the nation worse off than if we had remained a nation of farmers, fishermen, miners, and loggers. Just as that transition did not impoverish us in the past, there is no reason to believe that it is damaging us in the present.

CONCLUSION

Wildlands provide us with unique, non-reproducible "gifts of nature" that are increasingly scarce and therefore of increasing value. Commercial development of wildlands provides us with just another source of a commodity that could be produced at a variety of different sites and for which there are a variety of substitutes. Commodities, in general, are relatively plentiful and where they aren't we tend to find substitutes for them. That is the reason that most natural resource prices have been depressed for many decades.

It bears repeating:

Commercial development of wildlands amounts to the destruction of that which is unique and increasingly scarce for that which is common and replaceable with a small amount of effort. In a straightforward economic sense, this destruction of value makes no sense.

Preserving wildlands is unlikely to harm local economic vitality. Because people care where they live and act on those preferences, higher quality living environments also tend to attract both people and economic activity. Changes in our economy and in transportation and communication technologies have reduced the cost to people and firms of acting on their preferences for higher quality living environments. This has increased the economic importance of local quality of life in an area's economic base, and wildlands are a highly valued part of that quality of life.

In this economic context, wildland protection lays the basis for higher levels of current and future economic well-being. Rather than impoverishing our communities, wildland protection strengthens their current and future economic base and the sectors of the economy that will likely be the source of additional future jobs and income.

Editor's Note: The report that is the basis of this chapter can be viewed and downloaded from the Pew Wilderness Website at www.pewwildernesscenter.org.

QUANTIFYING LOCAL IMPACTS:
A MONTANA EXAMPLE
The Facts and Figures on the Economic Impact of Prohibiting
Logging in the Roadless Areas of Montana's National Forests

1. The Relative Importance of Forest Products in the Economy:
Most of Montana's forest products industry is located in western and
southwestern Montana. In the counties adjacent to the national forests
during the late 1990s, 3.4 percent of all jobs were directly
related to forest products activity (11,000 forest products jobs out of
340,000 total jobs).

**2. The Change in National Forest Timber Harvests and Local
Economic Impacts:** The peak year of harvest on both federal and
other lands was 1988. Since then the harvest from the national forests
has fallen by 76 percent, or 473 million board feet (mmbf). In the late
1990s, national forest land was the source of only about 15 percent of
total Montana harvests. During that same period the Montana counties adjacent to the national forests added about 91,000 new jobs.

**3. The Impact of Prohibiting Logging in Roadless Areas on the
Local Economy:**
Between 1993 and 1999 the national forest timber harvest that
came from the inventoried roadless areas in Montana's national
forests (NFs) was 4 million board feet (mmbf). For the 2000–2004
period, the Montana NFs plan to harvest about 3 mmbf in the roadless
areas. Thus, the "loss" to this part of the west central Montana economy of putting all roadless areas off limits to roading and timber
harvest would be a small reduction in NF timber harvests, at most 4
mmbf or about 2 percent. If these NF counties in Montana were able
to digest a 473 mmbf reduction during the 1990s without serious economic consequences, they should be able to digest a 4 mmbf reduction
with ease.
After a century of logging in Montana, the reason there are still
areas that have not been roaded and logged is that they are remote,

high-cost, and low-productivity areas where other forest values dominate. In the forest plans drawn up in the 1980s by the U.S. Forest Service 70 to 90 percent of the roadless areas in Montana's NFs were not included in the suitable timber base.

The direct employment impact of a 4 mmbf reduction in NF timber harvests would be about forty jobs. Each year during the 1988–1997 period about ten thousand jobs have been added in these counties. The forty job reduction represents about one job out of every 8,500 jobs in the Montana NF counties. Less than two days of normal job growth would replace them. Thus, ignoring any economic benefits of protecting these areas, the economic cost of permanent protection would be less than a one-hundredth of 1 percent reduction in employment or a two-day pause in the region's ongoing economic vitality.

Data sources: US BEA REIS CD-ROM for income; Region One U.S. Forest Service timber sale reports for timber harvests; U.S. Forest Service Roadless Initiative website for past and planned harvest in IRAs.

Between Towns and Wilderness: Protecting the Buffer Zones

Florence Williams

꙰ Lynn Owens lives and ranches along North Meadow Creek, a remote and sinuous drainage snaking down from the wild and partly roadless Tobacco Root Mountains in southwestern Montana. After following elaborate directions and several miles of dirt road, I find him, his wife Devonna, and a couple of friendly cow dogs in a modest white clapboard house with a spectacular view. He apologizes that the house is now harder to find, because he recently moved it a quarter mile off the dirt road.

I ask why he did that, and when he says "because of all the traffic," at first I think he is joking. He's yanking my chain because he knows I live in Helena, our small state capital with asphalt and stoplights and no wild, willow-lined cricks. But then I see that like a lot of Montana ranchers these days, Owens is dead serious. "There used to be a dozen cars a day on that road, and now there can be that many in five minutes," he says.

It turns out that Owens shares the road with two growing subdivisions just up the drainage, and each new resident seems compelled, he says, to drive into town (McCallister, population twenty-five) twice a day for mail and groceries. They live on twenty- or forty-acre ranchettes—some year-round, some just a few weeks a year. The dusty road is not the only thing making him queasy. He's also worried about the challenges of continuing to ranch next to a suburban neighborhood, where kids and dogs chase his cows and damage fences, where weeds spread unchecked, and where well-meaning but uninformed neighbors overdip their water rights.

Across the Intermountain West, it's a familiar story, and so is the sequel—ranchers with already marginal operations selling out, prime grassland lost to development, the landscape fragmenting and becoming denatured. In the last twenty years, half a million ranchers in the United

States have quit the business, and around the Owens place disaster stories aren't hard to find.

An hour to the northwest by car lies the Bitterroot Valley in Ravalli County. The local county commissioners have been averse to planning and zoning, and as Highway 93 winds eighty miles toward Missoula, it's lined with one- to twenty-acre lots, each containing a house. These subdivisions—even if relatively spread out—have helped to increase the county's population by a stunning 44 percent in just ten years.

An hour's drive in the other direction from the Owens ranch takes one to Paradise Valley, through which the Yellowstone River empties from Yellowstone National Park. With Gardiner at its head and Livingston at its foot, the valley is now home to exorbitant hobby ranches of wealthy out-of-staters, as well as a patchwork of suburban ranchettes.

Only twenty-five miles to the northeast sits the booming town of Bozeman, population thirty thousand, its expansion a source of dread for old-timers in still-rural landscapes nearby and with good reason: one of its bedroom communities, once-agricultural Belgrade, grew 67 percent to 5,700 people in the 1990s. This sort of growth shows up as high-density trailer parks, tract housing, and shopping malls. It also often results in stressed wildlife, damaged air and water quality, and overuse by recreationists on nearby public lands, like the Absaroka-Beartooth Wilderness and the unprotected roadless areas of the Gallatin National Forest.

But Owens and his neighbors in the sixty-mile-long Madison Valley northwest of Yellowstone National Park are determined to discredit the refrain that a rancher's last crop is a subdivision. Joined by conservationists, planning professionals, and some of the newcomers themselves, they have begun an ambitious, multi-pronged campaign to control their valley's fate and retain its rural and natural character. The Madison Valley Ranchlands Group, as it is now officially called, is one of many creative, collaborative conservation efforts emerging all over the West. Fueled by a mixture of good intentions, public and private funds, and a dose of desperation, these groups recognize the urgency of acting fast and employing a variety of tools, many still experimental, to preserve their way of life.

"Time is absolutely of the essence. Land values increase every year and only keep going up," says Jerry Scheid, an Idaho and Montana rancher who sits on the board of the Greater Yellowstone Coalition, a group that tries to protect private land surrounding the national park. Working ranches, in addition to providing economic and community stability, serve as critical buffer zones surrounding protected lands. These buffers—sometimes shaped like donut rings around the parks and wilderness areas, or like narrow corridors linking them—help safeguard everything from migrating wildlife to fragile hydrologic systems. Without these protective cushions, protected areas become "island wildernesses," their ecological and scenic values compromised by roads and the infrastructure that accompanies housing developments—shopping plazas, gas stations, restaurants, schools, and medical facilities.

Writing in *Conservation Biology* in 1998, Colorado State University ecologist Richard Knight noted that half of all threatened and endangered species depend on private land to survive. For example, even though 97 percent of Teton County, Wyoming, is publicly owned (an unusually high figure), the county's choice, lush, private lands provide habitat for 40 percent of the area's bald eagles, 80 percent of its trumpeter swans and 64 percent of its cutthroat trout, according to the Jackson Hole Conservation Alliance. Yet private open space continues to be lost every day. In Colorado, for instance, over 1.4 million acres of agricultural land were converted to suburban sprawl and other, non-farm uses between 1987 and 1997. Within the last decade, the average rate of loss there has doubled to over 270,000 acres annually. These figures will only accelerate. The 2000 census identified the inland West—which begins at the Rockies and contains most of the nation's unprotected roadless areas—as the fastest-growing region in America.

In remote Montana, land is still cheaper to protect than in other, more developed areas, and with the state's large ecosystems still functioning, the conservation bang for the buck is great. In fact, private-land conservation in Montana has far outstripped that of every other western state. Nearly one million acres of private lands in the state are permanently protected through development-restricting conservation easements, most of them on working ranches. Nonetheless, change is coming fast to these still-rural places.

For Owens, a conservative seventy-one year old, the sudden threats to his own backyard spurred him toward a level of activism that he never would have imagined back in the days when ranchers like himself pretty much minded their own business. But starting in the 1970s, the 3,300-acre spread next to the Owens place was transformed into the Shining Mountain Subdivision, platted with 125 twenty-acre homesites, about thirty of which have been built on so far. "I'd like to erase everyone of them damn things," says Owens, a lean, small man with snow-white hair and round glasses. Most of the lots sit right along Owens' north boundary. According to their deeds and covenants, each can be further split into five-acre parcels, for a potential build-out of five hundred houses. Then, in the mid-1990s, another developer appeared with plans to split a second three-thousand-acre ranch higher up the ridge into over a hundred small lots.

That's when Owens more or less got religion. He organized his neighbors and collected enough money for an attorney. Together, they filed an injunction against Madison County, which had approved the subdivision against the recommendation of its own planning board. Although the injunction expired, the developer agreed to alter his plans, creating fewer, larger lots. Says Owens, "The developer said to me, 'Okay, I'll do this, but you guys should be doing some planning of your own.' And he was right." In fact, Owens was so inspired that he helped found the Madison Valley Ranchlands Group. One of its first committees was the North Meadow Creek planning team.

"Zoning was a word flat outside of my vocabulary," recalls Owens, who raises Black Angus cattle. "I wouldn't touch it. Now, though, I feel some type of zoning is needed if we're going to have any control at all." Since he learned not to rely on the county for any planning authority, Owens and his neighbors have been working to form their own 25,000-acre zoning district, which the county will then be obligated to recognize. With the approval and signatures of 60 percent of the landowners in the district (out of about 180), they can write their own development codes. What they want is an exchange: if the Shining Mountain subdivision residents agree to only one additional split of their twenty-acre lots instead of three splits, the ranchers would agree not to subdivide their land to fewer than 80 or 160 acres. To protect larger parcels, some landowners have either placed or are considering placing conservation easements on their properties. In time, they could also sell or transfer development rights to appropriate, clustered building sites.

Here's how conservation easements work: pairing with a non-profit "easement holder," landowners voluntarily agree to restrict the way they use their property, usually by limiting future homesites, often on sensitive habitats like river and stream banks or elk winter range. The restrictions become a permanent part of the property's deed, and will remain in place for all future owners. In return for the property's lost development value, the rancher either receives cash or a tax break or a combination of both. Devaluation also helps relieve the burden of crippling estate taxes as ranches pass from one generation to the next. In addition, the cash infusion can help long-time ranchers pay debts and lease additional pasture to make more profits in an increasingly competitive commodities market. The non-profit enforces the easement and provides legal and technical assistance to the landowner.

In North Meadow Creek, most residents seem interested in some sort of restriction on development. "They built houses right next to my fence," explains Owens. "It was one of the [Shining Mountain's] selling points—you'll be right next to a big ranch where nobody will build in front of you. That's a toe-hold we have in the land-planning deal. They don't want someone building in front of them. It benefits them as well as us."

This area would be the first such zoning district in the county, but not the first in Montana. Gallatin County has sixteen such districts, including Springhill, which lies against the Bridger Mountains north of Belgrade. Owens considers it a model. There, fifty families on eight thousand acres agreed in 1992 never to subdivide beyond a density of one home per 160 acres. If an individual landowner wants additional homesites, he or she has the option to buy development rights from a neighbor, thus ensuring an overall development cap. Though the population of Gallatin County grew 34 percent between 1990 and 2000, an average of only one home per year has been built in the Springhill zoning district. "We figured if we could protect the ground in big enough chunks, agricultural land could still always be leased and thus ranching could be kept as a viable land use," says Springhill's Jim Madden. Some ranchers have willingly self-imposed even larger restrictions, such as the 640-acre minimum lots in the agricultural Boulder Valley south of Helena.

RANCHING AND WILDLIFE—LINKED FUTURES

Such large blocks of land are also useful to native flora and fauna, hence the interest of conservation-minded organizations in protecting ranches. For planning and political expertise during the past few years, the Madison Valley group has turned to the Sonoran Institute, a community-level, problem-solving spin-off of the World Wildlife Fund. In addition, The Nature Conservancy (TNC) will help the ranchers launch a program to purchase conservation easements in the area and will help fund a full-time staffer for the group. In the Madison Valley as a whole, some 76,000 acres are already permanently protected from development under deed-restricting easements held by The Nature Conservancy, the Rocky Mountain Elk Foundation (an organization that protects habitat and range land), a land trust called the Montana Land Reliance, and the state department of Fish, Wildlife and Parks.

Why are such ecologically minded organizations increasingly diverting precious funds to ranchers, whose methods and world views (often including predator control and riparian grazing) aren't always in line with their own?

"We don't unequivocally embrace all ranchers," says Jamie Williams, director of the Montana Nature Conservancy. "But we can, and should, reward the good stewards who live compatibly with these ecosystems. Keeping large, well-managed ranches intact is the best opportunity we have to protect vulnerable landscapes." As an example of compatibility, he cites the fact that far more grizzlies are killed on the heavily developed west side of Glacier National Park and the Bob Marshall Wilderness than on the east side, where sparsely populated land is still held in large ranches. "Never before has the fate of grizzlies and ranchers been so closely tied," he adds.

"We know that private ranch land is where a lot of wildlife habitat is," agrees Ray Rasker, director of the Sonoran Institute's Montana office. "The Madison Valley plays a critical role in the Yellowstone ecosystem. It's what links Yellowstone to the mountains that ultimately stretch to Glacier National Park. It's a key linkage corridor, with a lot of migrating animals using those private lands. If we lost the Madison to sprawl, then we've further insured that Yellowstone is just an isolated island, and then we lose animals at a faster rate." Wildlife that currently uses the valley for habitat and migration includes elk, moose, deer, wolverines, cougars, grizzlies, and wolves. The Madison Valley is surrounded by significant roadless and wilderness areas that these animals use, including the Tobacco Roots

Roadless Area to the north, the Red Rock Lakes Wilderness and Gravelly Mountain Roadless Area to the south, and the Lee Metcalf Wilderness to the east.

Rural residential development in the Montana and Wyoming portions of the Greater Yellowstone Ecosystem grew 400 percent between 1970 and 2000, according to Rasker. The population of Teton County surrounding Jackson Hole, Wyoming—which contains three wilderness areas—grew 107 percent. For bird species like yellow warblers, this was bad news, as development attracted more domestic predators, such as house cats, as well as natural predators, such as brown-headed cowbirds. The vast majority of bird "hot-spots" are found on lush, lower-elevation private land, not on public preserves. Furthermore, the authors said, residential development leads to increased fire suppression, road fragmentation, river channelization, sewage pollution, and the spread of exotic species like spotted knapweed and New Zealand mud snails, the latter two respecting no boundaries between developed and roadless lands.

Unchecked, these developments make the future look ominous. "We say that about one-third the land surrounding Yellowstone has already been platted into lots too small for agriculture, too big for a lawnmower," says Rasker.

The fact is, the conservation groups need the ranchers, and the ranchers need the funding, experience, and leverage the conservation groups can provide. Their cooperation extends beyond planning. To keep good agricultural stewards in business, Artemis Common Ground, a Helena-based non-profit, and The Nature Conservancy have helped develop a niche-marketing strategy in the Madison. In exchange for following a stewardship plan, transferring development rights either through a permanent or limited-term easement on their property, and agreeing to other standards (no added growth hormones, no feedlots), ranchers can produce Conservation Beef®. The dry-aged, grass-fed cuts are sold at a premium price to restaurants and private individuals, largely on the East and West coasts. Last year, its first, the program shipped 80 head to market; this year 150 head came from five ranchers. Such a program provides "long-term disincentives for subdivision" throughout the watershed, says Brian Kahn, director of Artemis Common Ground. "We hope to grow the program in the next phase on a much wider level."

This kind of "value-added" marketing—in which specialty goods can be sold for a premium—is slowly gaining popularity throughout western range lands, as ranchers struggle to increase their profits. The mother of all models is probably Oregon Country Beef, where Doc and Connie Hatfield organized their neighbors near the town of Brothers into a cooperative that sells natural beef directly to markets in Portland, Seattle, and San Francisco. Starting with fourteen producers and ten animals per week in 1986, the co-op now includes forty ranchers and sells 250 head each week for about $10 million annually. The ranchers are insulated from the price swings and corporate consolidations of the beef commodities market. "Niche marketing is working fantastic," says Connie Hatfield. "We'd be broke if we hadn't done this. Now we've built a sustainable organization, and we plan to be here for generations. Seven young ranch families have come back to the valley and say if it weren't for this they wouldn't be here."

BUILDING COALITIONS

In the ultra-scenic mountain West, with immigrants coming in from every part of the nation, demographic change is inevitable. Consequently, one challenge facing local communities and conservation groups alike is how to meld traditional ranch families and newcomers into a collaborative partnership. Once again, conservation easements seem to be the tool of choice.

In Montana, 90 percent of last year's land purchases of over one thousand acres were made by out-of-staters, and the Montana Land Reliance, perhaps the most successful state land trust in the country, has been able to place over 400,000 acres into conservation easements, most donated by wealthy buyers seeking charitable-contribution tax breaks. In the Madison Valley, conservation groups are working to leverage donated easements with additional private and federal funds. They then purchase easements from old-time ranchers who need cash to stay in business.

Such deals are already taking place in nearby Centennial Valley, located twenty miles west of West Yellowstone. To enter the spectacular fifty-mile-long valley is to step back in time one hundred years. About 90 percent of the private land—100,000 acres—is owned by only fifteen ranch families. All the ecological players and pieces are still there, from migratory birds to native trout to large predators, with the exception of free-roaming bison (although the occasional lone animal does wander through). Draining into the Red Rock Lakes National Wildlife Refuge and Red Rock Lakes Wilderness Area, the 6,500-foot-high valley houses the largest intact wetland system in the 18-million-acre Greater Yellowstone Ecosystem.

Only a few years ago, development in the valley seemed unstoppable, the shores of Henry's Lake, located just over the pass to the south, having already sprouted hundreds of vacation homes and lots, with new houses and paved roads appearing all the time. Yet for $12–15 million, local ranchers and The Nature Conservancy believe they can protect 45,000 critical acres, or nearly half of the entire private land. When private ranch properties are protected through conservation easements, leaving them economically and ecologically viable, so too are the thousands of acres of state and federal grazing leases that are tied to them. The leverage is financial as well: "If The Nature Conservancy can bring in a dollar of privately raised money, we can attract another four dollars worth of easement donation, and then match those five dollars in public money," says TNC's project director Tim Swanson, the former mayor of Bozeman and a consummate mediator. "The leverage we come up with is that one of our dollars leads to ten dollars of conservation. We need it because not every easement is donated. If we don't purchase some easements, we're going to lose the larger landscape."

So far in the Centennial, TNC has worked with private landowners and the U.S. Fish and Wildlife Service to secure over $7.5 million in public funding through the Land and Water Conservation Act ($3.75 million has been appropriated to date). In addition, they received $1 million from the North American Wetlands Conservation Act, of which $800,000 is appropriated for conservation easements. These public funds are being used to acquire 28,500 acres of conservation easements in the valley. Montana must rely on federal and private funds because the state, unlike many, is not currently funding a Purchase of Agricultural Conservation Easements program, and most local governments can't afford conservation on this scale.

As Garth Haugland, a self-styled curmudgeon and commissioner of Beaverhead County surrounding the isolated Centennial, says, "Our county can't—I'll be honest with you—do much with less than nine thousand residents, no toll station and no sales tax. How do we do [land protection] on a declining tax base? If [conservation] groups like this don't step forward, it won't happen. I guarantee you there's guys standing in line with a check waiting to buy these properties if you don't save them."

However, not all new landowners intend to subdivide. In the Madison, newcomers who were once eyed suspiciously have become integral players in the Madison Valley Ranchlands Group. Take Roger Lang, a Silicon Valley entrepreneur and venture capitalist who recently bought one of the largest ranches in the valley. A young part-time resident with a keen bent toward conservation, Lang quickly took an active interest in the weeds invading the valley. Noxious weeds have already spread to six million acres of Montana—private and federal lands, developed and roadless alike—and are a threat both to natural systems and to agricultural viability. "Because of the prevailing winds, my place had a reputation for being a weed seed bank," says Lang. Taking advantage of the locals' curiosity to see his house, previously owned by a prominent Hollywood mogul, Lang held an open-house fundraiser for combating weeds. With the proceeds, Lang and the Madison Ranchlands Group purchased two spraying machines for common use in the valley and Lang became the chair for the group's newest committee, which is charged with overseeing weeds. "Any way we can collaborate toward the benefit of ranchers is something I'm interested in," he says.

When one of Lang's ranching neighbors along the Madison River put his property on the market last year, Lang and The Nature Conservancy partnered to buy it, placed a development-restricting easement on it, and leased it back to the rancher, keeping him in business. "You should see the smile on that rancher's face," says John Crumley, a long-time rancher who now chairs the Madison Valley Ranchlands Group. "It took us a long time to realize that people other than us wanted to help us, that people who want to save our watershed don't have ulterior motives or some communist plot. We were once resentful of newcomers, but some of them are very impressive."

GRASS BANKS

To develop further partnerships with wealthy newcomers—many of whom do not actively ranch their own properties—the Madison group is putting together an experimental cooperative stewardship grazing program. The idea is that local ranchers can use pasture from the newcomers' land, as well as from each other's public lease lands, in order to rest, burn, or otherwise improve conditions on their own properties or public allotments. In return, the newcomers (and old-timers) get range-science expertise in managing their grazing lands and can enjoy a working partnership with their neighbors. As a result of cooperating and pooling their grass resources, ranchers can better manage grazing along natural boundaries and on a watershed scale, which ultimately means more grass for livestock and wildlife and more flexibility for ranchers to sustain economically viable operations. As examples of earlier cooperation, several ranchers jointly purchased a spring pasture, and one "conservation buyer" (a newcomer whose land is under easement) has designated his water rights for a common-good "instream flow," meaning it stays in the river to aid fish. He also provides grass on his twenty-thousand-acre property to neighboring ranchers for relief in drought years.

The snazzy term for a more formal version of this arrangement is "grass banking," which is one of the newest and most innovative conservation tools in the Western landscape. It was begun in the early 1990s by a coalition of southeastern Arizona and southwestern New Mexican ranchers called the Malpai Borderlands Group. The privately owned, 300,000-acre Gray Ranch is part of the group and leases out temporary grazing rights in exchange for ranchers placing conservation easements on their home properties. The easements are then held by the rancher-run, non-profit Animas Foundation. By resting their own range for up to several years, the ranchers can improve their land without selling off their base herd. Resting also allows enough bio-mass to accumulate for effective prescribed burns, as fire is a natural but long-neglected part of that grassland ecosystem.

Cooperating with the U.S. Forest Service and other agencies in the area, the Malpai Borderlands Group has successfully used fire to improve the productivity and ecological values of the range. The group has acquired easements on 35,000 acres so far; altogether, the project area encompasses over a million acres, the easements helping to protect it from the encroaching pressure of second-home development. This semi-desert grassland provides an important ecological bridge for wildlife migrating between the roadless areas of the Peloncillo Mountains to the southeast and the Chiricahua Wilderness to the northwest.

Such grass banks require a tremendous amount of funding, technical expertise, active land management, and neighborly and governmental coordination. Obviously, they won't work everywhere. But in large, threatened landscapes, they offer tremendous opportunity for motivated, creative partners. "Today, it would be no exaggeration to suggest that conservation grass banks have emerged in the West as the most promising high-leverage conservation tool to be devised in many years," says Bruce Runnels, the western regional director of The Nature Conservancy. In addition to areas in the Madison Valley, TNC is actively exploring grass banks in several states, including Wyoming's Absaroka Mountains, another critical ecological slice of Greater Yellowstone.

THE ROAD AHEAD

Huge challenges remain to conserving biologically important swaths of private land in the West. In places like Arizona, for example, where local communities have no subdivision zoning authority and where politically powerful developers eat up open space at an appalling rate, such work may depend on timely, careful strategies to purchase properties and development rights on the open market. Pima County surrounding Tucson is currently undertaking a major planning effort called the Sonoran Desert Conservation Plan, for example. It calls in part for buying easements on ranches with grazing leases on biologically significant state lands. If those ranches fall to development, the state-owned lands may follow. Some 53,000 acres of state trust lands have already been reclassified from grazing to commercial properties.

With funding from an open-space bond issue, the city of Tucson in 1998 bought the A-7 Ranch on the San Pedro River, one of the last intact perennial streams left in a critical wildlife corridor between the Rincon Wilderness to the west and the roadless areas of the Dragoon Mountains to the east. Through fire, rotational grazing, and the introduction of a grass bank for other ranchers on the ditch, the city hopes to help restore and protect the important water system as well as secure open space and protect linkages to other protected lands for wildlife. The 6,800-acre ranch leases 34,000 acres of state lease lands, which are now more likely to remain in grazing rather than being turned into commercial developments.

Hoping to replicate such projects, Pima County planners have identified priority watersheds and landscapes most threatened by Tucson's sprawl. "The most intact, roadless areas are the range lands," notes planner Linda Mayro. "But that's also where the developers want to go because it's cheaper," she says. The county faces the threat of "wild-cat" subdivisions, where speculators are allowed to divide parcels five times without review, then another five times, and so on, "until you're on one trailer per acre," says Mayro. In places like the San Pedro Valley, such a density would quickly deplete the water table, since a single person in Sonoita, Arizona consumes about ten times as much water as a cow.

But the county's top-down planning approach has already drawn fire from developers still flush from defeating the state's proposed growth plan in November 2000. The lesson seems clear: A grassroots approach—in which the multiple voices of a community articulate a unified goal—is often more successful than a government-initiated one.

Consider the Elk River Valley near Steamboat Springs, Colorado, surrounded by the Mount Zirkel Wilderness Area. Here both strong leadership and creative funding emerged in the mid-1990s, creating a model that has since widely influenced other parts of the West, including the Madison. One of the principal voices for conservation there is, ironically, the son of the founder of the Steamboat Ski Area. As property prices and growth spiraled out of control, Jay Fetcher saw his future as a rancher threatened by escalating estate taxes and encroaching sprawl. He and his neighbors wrote the "Upper Elk Valley Compact," a non-binding vision statement encouraging landowners to reserve a few low-impact building sites for future sales, as well as to place the rest of their property under conservation easements.

"We let the county know what we were doing and why," says Fetcher, who, with his father, runs a cow-calf operation. "We wanted to get their buy-in so that if someone proposed a mobile home park on an irrigated meadow, the commissioners would know that's not what we want." In fact, Routt County soon followed the Upper Elk by preparing its own "open lands" plan, with input from Fetcher and other ranchers, residents, and conservationists in the area.

Since the Fetchers set the tone by donating an easement across their 1,300-acre ranch to the American Farmland Trust (AFT), Routt County has exploded in easements: 34,000 acres as of 2001. Over 8,800 acres in Elk River Valley alone have been permanently protected by AFT and other trusts, including the local Yampa Valley Land Trust. This work is funded by a clever convergence of a voter-approved local property tax and the Great Outdoors Colorado (GOCO) lottery revenues program. Enacted statewide by voters in 1992, GOCO earns between $25–35 million a year and has so far protected more than 81,000 acres of Colorado's agricultural land through conservation easements.

"GOCO's partnership with agriculture is key to us achieving our mission of protecting river corridors, wildlife habitat, and open space," says John Hereford, GOCO's executive director: "Make no mistake, the road to conservation success runs right through the agricultural community."

It's a lesson the folks in the Madison Valley have taken to heart. There, conservation continues apace, spurred by rapidly changing times. "We figure we have about a ten-year window with baby boomers starting to retire and looking for affordable second homes," says Rock Ringling of the Montana Land Reliance. Hoping to head them off from building in the Madison River's floodplain, the group is eyeing two undeveloped subdivisions upriver of the booming town of Ennis. "We'll buy them, put an easement on them allowing one house per 320 acres instead of one per 20, and re-sell them," he says. If he sounds confident, he should be: the Montana Land Reliance already holds over forty-two conservation easements in the Madison Valley.

That's the kind of progress Lynn and Devonna Owens like to see. "I feel like we're coming along real good," says Lynn, who shares his ranch with deer, elk, coyotes, eagles, and perhaps the occasional wolf. "Give us a good living and we will put it back into the land in a way that it keeps going."

The Politics of Protecting Wild Places

Mike Matz

You kick back against a log, warmed and mesmerized by the crackling branches of a fire, and let your mind wander and ideas flow in the company of those who share your contentment with the cool night air, the stars in the sky, and the eerie shapes faintly illuminated around the edge of your campsite. The setting and your companions inspire a conversation about wild places, their importance to people and wildlife, and how some of these fast-vanishing landscapes might be saved, making the world a better place—or at least your corner of it. In the end, you have all the ingredients for concocting something great. You toss out a few suggestions. Your friends add something relevant or whimsical. The embers in the fire burn brighter, more wood is added, and a dream takes shape.

Something similar happened in 1934 along a Tennessee road when Alaska wilderness explorer Bob Marshall, the Appalachian trail's father, Benton MacKaye, and one of the backers of Great Smokies National Park, Harvey Broome, created The Wilderness Society. In the ensuing decades, the organization took the lead in promoting passage of the 1964 Wilderness Act, and played a crucial role in lobbying Congress to include parts of America's remaining wild places in the National Wilderness Preservation System. Each meeting of The Wilderness Society's governing council now honors its Tennessee conception by gathering around a bonfire and sharing thoughts about the state of our wilderness.

In 1997, another group of bright, young conservationists gathered around a campfire in Utah's Canyonlands National Park after a day of hiking amid spires of sandstone. Among them were Melyssa Watson and Brian O'Donnell. At the time, each was twenty-six years old, Watson working as a lobbyist for the National Family Planning and Reproductive Health

Association and inspired by a recent wilderness conference in Tucson she'd attended, and O'Donnell serving as the executive director of the Alaska Wilderness League and embroiled in the battle to save the Arctic National Wildlife Refuge from oil drilling. As they watched the flames, they pondered a problem—in short, that the wilderness protection movement had become mired in mud. Aside from a few places like Utah and Colorado with viable campaigns for designating wilderness, little was happening in other states, and Watson and O'Donnell decided that they needed to become a kind of SWAT team for wilderness protection. Their weapons would be grassroots organizing and media outreach, their chief tactic, training local conservationists in the art of using these weapons with political savvy. "I remember thinking, jeez, we *had* to carry on the work of those who did it before us," O'Donnell recalls, "if the wilderness movement was to be sustained."

The couple traveled around the country and sounded out other conservationists on their SWAT-team idea. Many, including William H. Meadows, president of The Wilderness Society, and Bert Fingerhut, chairman of its governing council, liked what they heard. The two men thought highly enough of the pair's assessment and proposal to take the matter to the organization's governing council for adoption, and in May 1999, under the aegis of The Wilderness Society, the idea for a SWAT team became the Wilderness Support Center based in Durango, Colorado. Some dreams plotted around campfires dissipate like smoke; this one became a reality.

Efforts of the Wilderness Support Center first bore fruit in Nevada, a state where 83 percent of the land, 60 million acres, is held in trust by the federal government for every U.S. citizen—people in New Jersey as well as those in Nevada. Of course, not all of Nevada's federal lands qualify for wilderness designation. Some of them have been mined, others have been peppered with military communication towers, some of the state's rivers have been dammed to create reservoirs over once untamed land. Any of these imprints of civilization disqualify a piece of otherwise wild country for protection under the 1964 Wilderness Act,[1] which stipulates that potential wilderness lands be affected primarily by the forces of nature. Nevertheless, millions of acres of Nevada's basins and ranges do qualify for protection as legal wilderness, even though the state had only 800,000 acres of designated wilderness in 1999. By way of comparison, the very same acreage had been set aside in Minnesota and Florida, even though these states were settled more than a hundred years before Nevada and have far less public land from which to choose potential wilderness sites.

"Nevada epitomizes the vast and unoccupied landscape that the rest of America used to be," said John Wallin, director of the Nevada Wilderness Project, a Reno-based group that works hand-in-glove with the Wilderness Support Center. "Plus, it's the most mountainous state, with more than 350 mountain ranges. Nevada has unbelievable potential for wilderness, and that's what I like about it."

The first step Wallin took to gain protection for some of Nevada's wild places was to survey them—a daunting task that he began in 1999. The thirty-five-year-old conservationist had honed his organizational skills as a manager of Patagonia's mail order warehouse in Reno; now he needed to put together a cast of volunteers, train them to use standard criteria to mark up their maps, verify the field work, and assemble the data into a statewide assessment of what wilderness remained before putting together a citizens' proposal for new wilderness areas. But an unlikely turn of events threw him off course.

Senator Richard Bryan of Nevada planned to retire at the end of 2000 and had begun to turn his thoughts to his legacy, just as President Clinton was doing during his last year in office when he signed the Forest Service's Roadless Conservation Rule, a regulation that precluded any new road construction on 58.5 million acres of our national forests—an amount of real estate larger than Kansas.[2] President Clinton also had proclaimed about two dozen new national monuments encompassing over ten million acres under authority granted him by the 1906 Antiquities Act.[3] The 1.9-million-acre Grand Staircase-Escalante National Monument, connecting Glen Canyon National Recreation Area with Bryce Canyon National Park in remote reaches of south-central Utah, became the first established by Clinton, in 1996.

Watching this flurry of presidential action to conserve some of our disappearing wild heritage, Senator Bryan concluded that the Black Rock Desert in northwestern Nevada—filled with pronghorn antelope, mountain lions, and the wagon ruts of pioneers—merited similar consideration. In March 2000, he introduced the Black Rock Desert-High Rock Canyon Emigrant Trails National Conservation Area Act. But its prospects for passage were poor: little time remained in that Congress' second session for committee hearings and floor action on both sides of Capitol Hill.

The Wilderness Support Center threw all its efforts into getting the act passed, O'Donnell camping out with Wallin in a cheap motel on the outskirts of Washington to watch and perhaps to influence the course of last-minute, late-hour wrangling that occurs in the crush of congressional business during the month before adjournment. Watson joined them for a couple of weeks. The three called Senator Bryan's staff almost every hour to get information that would help to facilitate supportive participation by other members of Congress and their staffs. Watson then flew to Nevada to assist in coordinating involvement of volunteers back in the state, whose letters to editors streamed into Las Vegas and Reno papers in support of the bill.

"The volunteers were just great," recalled Watson. "They would go to a breakfast sponsored by Bryan and provide him encouragement along with his pancakes."

The hard work of O'Donnell, Wallin, and Watson paid off. Assisted by Nevada's other senator, Harry Reid, who joined forces with Bryan and lent a helping hand in the final hours of Congress, Bryan passed the Black Rock Desert-High Rock Canyons Act, which established a 1.2-million-acre national conservation area within whose boundaries were 740,000 acres of officially designated wilderness, nearly doubling the amount of Nevada's wild landscape protected in perpetuity.

"We really challenged conventional wisdom in this state," Wallin added. "We showed people that we could take command of our own destiny and that those who oppose safeguarding our public lands can't any longer just dig in their heels. We also showed people how to have a lot of fun."

The Nevada Wilderness Project and their colleagues at Friends of Nevada Wilderness in Las Vegas are hardly finished with protecting the state's wildness. In partnership with the Wilderness Support Center, the two groups are now working with Senator Reid to fashion an omnibus lands bill that could designate as much as four million acres of wilderness in the starkly beautiful Mojave Desert. The bill would be an important step in maintaining some degree of open space in one the nation's fastest growing regions.[4] Since Senator Reid helped persuade Senator Jim Jeffords to jump from the Republican Party and join the Democratic caucus as an independent (a switch that handed over control of the Senate to the Democrats), he rides high in esteem among some of his colleagues, increasing the odds that this omnibus bill will become law.

LIGHT AT THE END OF THE BUSH TUNNEL

Such measures for the protection of our nation's dwindling supply of natural landscape—especially in a western, conservative state like Nevada—are indeed signs of hope that the National Wilderness Preservation System will continue to expand. However, no conservationist can rest easy with George W. Bush in the White House. Not only is Bush philosophically opposed to wilderness conservation, he is also indebted to the politicians of the Rocky Mountain states who helped to elect him. These conservative members of Congress happily line up for federal subsidies to underwrite the costs of logging and dam-building on their home turf, but generally oppose federal rules or laws that protect those same wide-open spaces.

To assuage this constituency, President Bush appointed Gale Norton as Secretary of Interior and Ann Veneman as his Agriculture Secretary. As both a protégé of James Watt, and as Colorado's attorney general, Norton has often criticized and worked to dismantle environmental laws.[5] Veneman, who now oversees the Forest Service, most recently acted as a legal representative for both agribusiness and biotechnology firms, where she opposed regulating these industries.[6]

Further catering to his western constituency, President Bush suspended implementation of the Forest Service's Roadless Conservation Rule and effectively reopened the 58.5 million acres that had been protected from new road construction. Ironically, the Forest Service, which already manages 386,000 miles of roads on its lands (more than our entire interstate highway system), has stated that revising this forest conservation policy was necessary in order to involve the public in the decision. The federal agency has conveniently forgotten the six hundred public hearings it conducted during 1998 and 1999, which generated more than 1.6 million comments, 90 percent of them in favor of the policy.

"What the administration really wants is a way to kill a needed preservation rule approved during the Clinton presidency that its timber, mining and oil buddies don't like," opined an editorial in the *Atlanta Constitution*.[7] The rule's suspension for ninety days bought time for timber interests and for the governors in Idaho and Alaska to challenge the rule in court. Eventually Idaho found a sympathetic judge who enjoined the policy from taking effect.[8] Meanwhile, the new chief forester has opened the way to build logging roads on 11 million of the 58.5 million acres covered by the Roadless Conservation Rule.

Speaking of Bush and his initial policy decisions, Democratic pollster Peter Hart said, "Here's a person who came into office with a wide-open playing field, and within six months, he's had the ability to align himself with two major forces, one being the right of the Republican Party and the other being the special interests of America."[9]

There could well be a brighter side to this outlook for conservation of public lands. President Bush saw his standing in opinion polls drop early in his term because of his decision to abandon a campaign pledge for tighter emissions of carbon dioxide, his willingness to permit higher levels of arsenic in drinking water, and his rejection of a global treaty on climate change. In March of 2001, 36 percent of the adults surveyed in a CNN/Time poll thought the president was doing a poor job in handling environmental issues;[10] the following month his environmental disapproval rating had risen to 41 percent;[11] and by August, 49 percent of the American public didn't like the way he handled environmental issues.[12] A still higher 53 percent disliked his plan to address the nation's energy needs—a plan that focuses heavily on increasing production by drilling holes into public lands while doing little to conserve energy.[13]

Clearly, Americans don't seem to agree with their president's environmental stances, but as Republican pollster Linda Divall remarked, "He wants to maintain his conservative coalition. At the same time, if he is going to be a successful president, he has to build centrist and bipartisan support."[14] In building such support and angling for a second term, President Bush will have to bolster his image with the swing moderates in suburban America for whom environmental issues are front-and-center. To do so he may label his policies "compassionate conservatism," yet he'll stray very little from the path of the far right, his bureaucrats, newly ensconced inside the walls of the Interior and Agriculture Departments, working quietly to scuttle wilderness recommendations as they testify in opposition to bills before Congress.

Fortunately, Congress alone designates wilderness for permanent protection, and given its present even division—Republicans holding a slim twelve-seat advantage in the House of Representatives and Democrats clinging to a one-seat margin in the Senate—Congress could send, if pressured by concerned voters, a series of wilderness bills to the president for his signature. Acutely aware of the need to bolster his environmental record, Bush will more than likely sign them. The proof for this prediction? None other than President Ronald Reagan. The standard-bearer of the far right signed twenty wilderness bills in 1984, more bills in one year than any president since the 1964 Wilderness Act was enacted. Reagan, who appointed James Watt as his first Interior Secretary, was no conservation champion, yet he added over eight million acres to the National Wilderness Preservation System. Congress sent him bills to sign and, except for one case, Montana, he found no plausible cause to refuse these members what their constituents back home wanted them to deliver.

PROSPECTS AND PLAYERS

According to a *Los Angeles Times* poll, nine of ten Americans say that wilderness is important to protect,[15] and to gain reelection congressmen and senators must listen to these constituents as much as to the special interests that heavily fund their political campaigns. Consequently, they consider proposals for creating more wilderness that are formulated by constituents alongside recommendations from the Forest Service and the Bureau of Land Management (BLM), the two agencies that hold much of the nation's wild places in public trust, but which are often influenced by the oil, gas, mining, and timber industries.

A spectacular example of this political dynamic has taken place in Utah, where the BLM recommended only about two million acres for inclusion into the National Wilderness Preservation System while the state's residents identified over nine million BLM acres that are suitable. The residents' recommendation is pending legislation while the agency's isn't. A similar pattern of the electorate's influence on what will become wilderness can be seen across the country. Starting in the West, where the lion's share of natural heritage remains, here's a sample of what the conservation community and its allies in Congress might add to the National Wilderness

Preservation System in the next few years. California is a good place to start because it illustrates well the convoluted horse-trading that goes into passing wilderness bills.

The California Wilderness Coalition has identified what wilderness remains in the nation's most populous state and is now making Californians aware of their last wild places. In the meantime, Senator Barbara Boxer of California has told conservationists that she intends to introduce a wilderness bill for between two and four million acres of publicly owned land across her state after talking with local elected officials and county commissioners in affected areas. In the House of Representatives, members like Mike Thompson of northern coastal California and Lois Capps, who represents the south-central coast north of Los Angeles, may introduce their own bills for designating wilderness, as Representative Sam Farr has indicated he will do for his central coast district.

With Democratic Senator Jeff Bingaman of New Mexico now chairing the Senate Energy and Natural Resources Committee, any legislation Democratic Senator Boxer eventually authors is likely to get a fair hearing and have a respectable shot at passing through committee and getting onto the floor for consideration. But Senator Boxer must first come to agreement with her California colleague Senator Dianne Feinstein on her statewide wilderness legislation.

The Senate—not always, but often—defers to the wishes of the members from the state a bill most affects, so if there's consensus between the two Democrats, a bill for increasing California's protected land could well reach the president's desk sometime during the 107th Congress. Given the number of California's electoral votes, and Bush's desire to see if he can get them into his column in 2004, he probably won't veto the legislation.

In Colorado—a state that did go to Bush in 2000—Representatives Mark Udall and Diane DeGette have both introduced wilderness bills. Udall's designates wilderness across three counties. In two of them, county commissions support his bill; but in Grand County, opposition to the bill prompted Udall to create an 18,000-acre "special protection area" requiring the Forest Service to manage rampant off-road vehicle use more conscientiously and to preclude any sale or trade of the land in the area. Conservationists dislike this sort of ersatz wilderness, for it doesn't prevent dirt bikes and four-wheelers crisscrossing the area. "We're pushing Udall to make the whole area wilderness, "said Jeff Widen, co-chair of the Colorado Wilderness Network.

DeGette's legislation mirrors a citizens' proposal that would protect 1.5 million acres of canyon country on the western side of the Rocky Mountains, though none of this country lies in her Denver metropolitan district. It's in the district of Representative Scott McInnis, who chairs the House Resources Committee. He considers Degette's measure too sweeping, and he can and probably will foreclose any action on it.

Yet, McInnis doesn't mind plucking the most scenic and least controversial places from DeGette's package, as he did during the 106th Congress in 2000 while the DeGette's bill stalled. He worked to make her proposal for such popular destinations as Black Ridge, Gunnison Gorge, and Spanish Peaks part of the National Wilderness Preservation System. In the 107th Congress, he's also reported to be looking at sponsoring a bill to protect the 22,000-acre Deep Creek Canyon, which is also contained in a citizens' proposal and for which McInnis has an affinity since as a youngster he hiked and camped there. Whether his constituents can convince him to include all of the 22,000 acres they've proposed is unknown.

Utah and Idaho typify the flip side of wildlands conservation—the wilderness advocates in both states often need to make lemonade out of political lemons. For instance, Utah Representative James Hansen, who became chairman of the full House Resources Committee in the 107th Congress, has the same problem as McInnis: increasing national support for protecting redrock canyons and basin-and-range country in Utah has put pressure on him to accede to the wishes of wilderness advocates, whose call for protection of nine million acres is embodied in legislation put forward by Representative Maurice Hinchey of New York and 151 of his colleagues. But Hansen, an ardent wilderness foe, has taken a much harder line than McInnis.

Hansen wrote a bill in the first session of the 107th Congress for the Pilot Peak Range near the border with Nevada that will allow unprecedented motorized access as well as construction of military buildings and communication facilities, both of which contradict the meaning of wilderness as defined by the 1964 Wilderness Act. He has also included only half of the fifty thousand acres in the Pilot Peak Range that were identified by citizens in Utah as suitable for wilderness designation. For conservationists, the Hansen bill is a bitter pill to swallow. As an antidote, conservationists hope for backing from Representative Jim Matheson, a middle-of-the-road politician who represents an urban Salt Lake County where support for wilderness protection runs deep.

But relying upon Matheson for help makes people like Larry Young, executive director of the Southern Utah Wilderness Alliance, nervous, since the Utah representative hasn't been willing to take a stand on wilderness issues. Young, a descendent of the Mormon leader Brigham Young and a former professor at Brigham Young University in Provo, testified at a hearing on the bill before Hansen's committee in the 107th Congress, and later said, "If we hold onto Matheson, we should be able to eliminate the bad parts of the bill and improve on the amount of acreage it would protect. That'll make the bill palatable to us."

In Idaho, wilderness allies are also trying to turn a defensive campaign into an offensive one. The U.S. Air Force has maintained that a remote corner of southern Idaho known as Owyee Canyon would make an ideal bombing range; but this vast sweep of high desert and gorges is also a haven for bighorn sheep, for the increasingly rare sage grouse, and for people looking for a bit of solitude. Conservationists unsuccessfully urged the Clinton Administration to proclaim Owyee Canyon a national monument.

The push wasn't for naught, though. People who didn't have a clue about the region's beauty now not only know about its wild qualities but also have a sense of the widespread support for protecting the canyon and its side channels. The 10,500 inhabitants in the vicinity recognize that changes are coming to their corner of Idaho as more and more people move or travel to the state, and they have begun to see wilderness designation as a means to protect their rural way of life. Senator Michael Crapo, not unaware of the demographic changes affecting Idaho, is thus fostering collaboration among county commissioners, ranchers, and conservationists in what's become known as the Owyee Initiative. If Senator Crapo can figure out how to "keep the parties negotiating with each other, not talking at each other," said an editorial in the Idaho Statesman, he could put together a proposal that "has the makings of a political legacy."[16]

Senator Crapo may also be willing to sponsor a wilderness bill for the Boulder-White Clouds, a breathtaking mountain range situated in central Idaho that has been long promoted for protection. At a half million acres, with headwaters of four wild rivers, this may be the largest chunk of wilderness left in the lower forty-eight states. If Crapo does sponsor a bill for the area, wilderness nemesis and fellow Idaho Senator Larry Craig may well follow along. He told a group of sportsmen who visited him in Washington, D.C. that he would support wilderness designations.

Commenting on the changing face of wilderness politics in Idaho, Rick Johnson, executive director of the Idaho Conservation League, said, "We're helping to make sure people here are aware of their natural heritage, and in turn they're helping our members of Congress identify with these particular places. When that happens wilderness protection is politically inevitable even in this conservative place. Idaho, is, after all, the wilderness state."

Despite Johnson's contention, Alaska has an even greater reservoir of wilderness—58 million acres of it already safely tucked away in the National Wilderness Preservation System. For the state known as "the last frontier," Representative Rosa DeLauro of Connecticut intends to sponsor a wilderness bill for Chugach and Tongass National Forests of about 14 million acres. Conservationists from the Alaska Coalition are also turning their eyes toward a BLM wilderness review of Alaska's public lands—a pool of 86.5 million acres of mostly unspoiled terrain. They also have to defend the Arctic Refuge against an onslaught by the Bush Administration, the oil industry, and a few labor unions. "Our chances are certainly better in the Senate than in the House to protect the Refuge," said Adam Kolton of the Alaska Wilderness League, a catalyst among the 160 groups comprising the Alaska Coalition, "but none of us should be under any illusion that this fight will be easy."

Never easy, never fast, is indeed how these conservation battles go, no more so than east of the Mississippi, where the amount of land left in a wild state is much smaller than in the West. Nonetheless, these eastern lands provide important wildlife habitat and recreational opportunities. In Pennsylvania, a tiny group called Friends of the Allegheny is trying to designate some of the last old-growth forest in a natural area near the town of Tionesta, the birthplace of the 1964 Wilderness Act's author, Howard Zahniser. New Hampshire's White Mountain National Forest sees more visitors annually than Yellowstone and Yosemite National Parks combined and local residents are promoting new wilderness additions there. Efforts to jump start wilderness campaigns are also underway in Alabama, Georgia, North Carolina, Tennessee, Vermont, Virginia, and West Virginia.

Yet for all the hullabaloo within the conservation community and in the corridors of Congress, many Americans remain astonishingly unaware that Congress has a tool, in the shape of the 1964 Wilderness Act, to protect open space, safeguard air and water quality, and save habitat for wildlife. While seven of ten people say they've been to a national park,[17] only five in ten know the National Wilderness Preservation System exists and only three in ten know for certain that they've ever visited one of the 630 places that are included in the NWPS.[18]

This lack of knowledge on the part of the public about wilderness was one of the very reasons that the Pew Wilderness Center was established in April 2000. To complement efforts of other wilderness groups in raising public awareness, it conducts education initiatives on television, radio, and in newspapers, all aimed at alerting Americans that their nation's remaining wild places can be saved before they're lost to shopping malls and housing tracts.

A key partner in this endeavor is the American Wilderness Coalition, created in 2001 to show citizens how they can contact their members of Congress and, in essence, lobby for the preservation of wilderness. Another player on the scene is a political action committee, WildPAC, specifically set up to support champions and supporters of wilderness in Congress. Outdoor recreation companies like Eagle Creek, Patagonia, and REI have also become involved through an organization called Businesses for Wilderness, which helps these businesses to participate in campaigns to save wild places. Their bottom-line rationale is self-evident: the less wilderness, the fewer the opportunities for people now and in the future to visit these areas and use outdoor products.

Wilderness bills are sprouting like glacier lilies in the spring, and the conservation community, better equipped with resources and newly formed alliances, is poised to help enact them into laws that afford permanent protection. A success such as the creation of the Grand Staircase-Escalante National Monument in 1996 breeds others, like the Black Rock Desert-High Rock Canyons National Conservation Area. A growing list of citizens from all walks of life, aided by the likes of the Wilderness Support Center's Brian O'Donnell and Melyssa Watson,

Nevada's John Wallin, Utah's Larry Young, and Idaho's Rick Johnson, are carrying on the tradition of those who dreamed around campfires, hoping to protect wild places, turning those dreams into reality, and saving their corner of the world.

Christianity and Wild Places

Steven Bouma-Prediger

∾ Many people blame Christianity for our present ecological crisis and argue that Christian faith legitimizes the exploitation of both land and wildlife. In short, say these critics, the term "Christian environmentalist" is an oxymoron.

Given the opening chapters of Genesis, it's no surprise how this view came about. After creating earth and heaven, waters and wildlife, God creates man and woman, and tells them: "Be fruitful and multiply, and fill the earth and subdue it; and have dominion over the fish of the sea and over the birds of the air and over every living thing that moves upon the earth." (Genesis 1:28). As the American historian Roderick Nash points out, "It followed that the Christian tradition could understand Genesis 1:28 as a divine commandment to conquer every part of nature and make it humankind's slave."[1]

Wed this several-thousand-year-old Judeo-Christian tradition of dominion over the earth with the rise of technology in the Middle Ages and a true anti-ecological juggernaut was born. This is the central thesis of Lynn White, Jr., who taught medieval history at UCLA. His famous 1966 essay, "The Historical Roots of our Ecological Crisis,"[2] makes the claim that "Christianity bears a huge burden of guilt"[3] for the present eco-crisis because of its underlying role in the rise of modern Western science and technology.

Since God created nature, goes White's argument, the mind of the divine can be revealed to humanity through the empirical investigations of science. Furthermore, White argues, by emphasizing both divine and human transcendence over nature, Christianity fostered indifference toward the natural world and thereby sanctioned its exploitation. Thus, by means of their knowledge of and control over the world—that is, through science and technology—humans could exert unprecedented power over nature with a clear conscience.

"Christianity," notes White, "in absolute contrast to ancient paganism and Asia's religions (except, perhaps, Zoroastrianism), not only established a dualism of man and nature but also insisted that it is God's will that man exploit nature for his proper ends."[4] The results are the ills with which we're all familiar: loss of grasslands and forests; extinction of species; toxic waste; global warming.

But is the Bible really as anti-ecological as White and his followers suggest? Do the first two chapters of Genesis actually license the exploitation of creation? Many Christian scholars, myself included, believe that the answer is no, and we base our assertion on how the original Hebrew meanings of dominion, and the contextual backdrop of the rest of the Genesis story, have been lost in translation.

The opening of Genesis does indeed indicate that one dimension of humanity's calling is mastery. In the Hebrew, the human is asked to subdue (*kâbâsh*) and have dominion over (*râdâh*) other creatures. But what exactly is the shape of this dominion? The surrounding context of Genesis 1-2 (*New Revised Standard Version*) sheds light on this important question.

For example, Gen. 2:5—"and there was no one to till the ground"— literally speaks of there being no human earth-creature to serve the earth (no *'adâm* to *'abâd* the *'adâmâh*). Gen. 2:7 makes clear that the human earth-creature (*'adâm*) has the name it does precisely because it is made from the earth (*'adâmâh*): "Then the LORD God formed man of the dust of the ground, and breathed into his nostrils the breath of life; and the man became a living being." To carry the Hebrew word-play into Latin, we are humans because we are from the humus. Therefore, other earthly creatures, also made from the ground, are our sisters and brothers.

Furthermore, Gen. 2:15—part of which is quoted on the door of every Chicago police car, "And the LORD God took the man, and put him in the garden of Eden to serve it and protect it"—defines the human calling in terms of service. We are to serve (*'abâd*) and protect (*shâmâr*) the garden that is the earth for its own good, as well as for our benefit. Dominion is thus defined in terms of service, not domination; to focus only on certain texts and then to interpret them as necessarily entailing domination is faulty exegesis of the first order. Only a selective and biased reading of Genesis 1-2 could lead to the abuse of our planet.

This understanding of dominion as service is confirmed when other and often-neglected texts in the Bible are taken into account. Consider Psalm 72, which speaks most clearly of the ideal king, one who rules and exercises dominion properly: "For he delivers the needy when they call, the poor and those who have no helper. He has pity on the weak and the needy, and saves the lives of the needy. From oppression and violence he redeems their life, and precious is their blood in his sight." The psalm unequivocally states that one who rules rightly executes justice for the oppressed, delivers the needy, helps the poor, and embodies righteousness in all he does. In short, the proper exercise of dominion yields *shalom*—that pregnant Hebrew term which means the flourishing of all creation. This is a far cry from dominion as domination.

In the New Testament, Jesus also defines dominion in terms other than exploitation. For Jesus, to rule is to serve (Matthew 20:25-28): "But Jesus called to them and said, 'You know that the rulers of the Gentiles lord it over them, and their great ones are tyrants over them. It will not be so among you; but whoever wishes to be great among you must be your servant, and whoever wishes to be first among you must be your slave; just as the Son of Man came not to be served but to serve, and to give his life a ransom for many.'" To exercise dominion is to suffer for the good of the other, if necessary like Jesus to the point of a cross (John 13:34): "I give you a new commandment, that you love one another. Just as I have loved you, you also should love one another." So although humans are called to rule, the ruling must be understood as a form of service.

Essayist, poet, and Kentucky farmer Wendell Berry puts this well:

> Such a reading of Genesis 1:28 is contradicted by virtually all the rest of the Bible, as many people by now have pointed out. The ecological teaching of the Bible is simply inescapable: God made the world because he wanted it made. He thinks the world is good, and He loves it. It is His world; He has never relinquished title to it. And He has never revoked the conditions, bearing on His gift to us of the use of it, that oblige us to take excellent care of it. If God loves the world, then how might any person of faith be excused for

not loving it or justified in destroying it?[5]

Yet another common environmental criticism of the Bible has more to do with what will happen at the end of the world than how things were meant to be in the beginning. Many argue that Christian eschatology undercuts any reason for preserving the earth since the return of Jesus will usher in a completely new age. In effect, the second coming militates against our caring for the earth since this world is ephemeral and ultimately unimportant.

A famous example of someone who represents this perspective is James Watt, the first Secretary of the Interior under President Ronald Reagan. In response to a question from a congressman, about why the Department of Interior, contrary to its expressed mandate, was not taking better care of our national parks, Watt, a devout Christian, said, "I do not know how many future generations we can count on before the Lord returns."[6] Like many other Christians, Watt believes that because everything will be destroyed when Jesus returns, there is no need to care about Yosemite or Yellowstone, Isle Royale or the Everglades. This "pervasive otherworldliness of Christianity," according to Roderick Nash, has fostered a view of the earth as "a kind of halfway house of trial and testing from which one was released at death."[7]

But is such an eschatology biblically accurate? Will the earth be destroyed after the Last Judgment? The answer, again, is no. A properly biblical view of the future does not envision the destruction of the earth, but rather its redemption, restoration, and transfiguration. God's good future is not other-worldly but this-worldly. It is earth-affirming, not earth-denying. For example, 2 Peter 3:10, when properly translated, declares that the new earth will be found, not burned up—discovered, not destroyed. The Greek verb in this passage is *heurêthêsetai*—from *heureskein* or "to find" and from which we get the English expression "Eureka." So the correct translation of the last clause (found only in the New Revised Standard Version and the Dutch *Niewe Vertaling* 1975) is "and the earth and everything that is done on it will be found."[8] Properly understood the Bible does not teach that the earth is ephemeral and unimportant. Rather, it is full of crystalline rivers and fruiting trees, a heaven on earth, as described in Revelation 21-22. God's good future is not a world of all new things, but a world in which all things are made new.

But even if Watt and others were right about the destruction of the earth, why does it necessarily follow that we should not care for it now? It is a non sequitur to argue that simply because the earth will be destroyed in the future, humans should exploit it in the present. Because something will eventually be destroyed gives no license for abuse or neglect. No housekeeper would ever suggest that because his or her tenure were temporary, the house should not be maintained.

What then are some compelling reasons to take care of the larger house of our wildlands and wildlife?

First and foremost, and no matter our religious beliefs, humans ought to care for nature because everything is connected to everything else. Aldo Leopold summarized this neatly when he said that "all ethics rest upon a single premise: that the individual is a member of a community of interdependent parts."[9] The preservation of wildlands is important not just because it provides places of recreation and respite for humans, but because preserving such land contributes to the health of the entire biotic community, including ourselves.

We should also care about wildlands and wildlife—the Arctic Refuge and caribou, the Everglades and herons—because they have value over and above any usefulness to humans. They have what ethicists call intrinsic value. Christians certainly have good grounds for accepting this argument. For example, Psalm 104 insists that non-human creatures have worth irrespective of their usefulness to us. The mountains are important for wild goats, the cedars are needed by storks, and the seas support Leviathan. As Psalms 96 and 148 trumpet, all creatures are designed to sing praises to God. They have value as part of the grand symphony that is creation. Contemporary Reformed theologian Jürgen Moltmann reiterates this claim when he says, "The creatures of the natural world are not there for the sake of human beings. Human beings are there for the sake of the glory of God, which the whole community of creation extols."[10]

Moltmann's view displaces humans from the center of the universe and reminds us that wildlands and wildlife are valuable to God, and therefore should be valuable to us. As the Hebrew prophets remind us, the trees and rivers are able to praise God. To see a tree as only so many board feet, or a river as only a place to fish, is a form of myopic utilitarianism which reduces all value to human terms. A focus only on human use—even if wise use—is a stunted viewpoint which fails to acknowledge value in a world not of our making. In sum, we have duties to care for non-human creatures for their own good, as well for the goods we acquire from them.

People of all faiths can certainly embrace these principles of right behavior toward wildlands. Christians, in addition, have other compelling reasons to care for the Earth.

First and foremost, we must honor our planet's wild places because God asks us to. Authentic faith demands that we obey God, and God commands that we be Earthkeepers—"The LORD God took the man, and put him in the garden of Eden to till it and keep it." (Genesis 2:15)

Second, since in the Christian view humans are meant to be God's image-bearers, and since being an image-bearer of God involves caring for the needs of others, Christians are called to show the kind of care that God exhibits. God's concerns are our concerns. The classic scriptural warrant is Genesis 1:26-27: "And God said, Let us make humankind in our image, according to our likeness." We are created in the image of God. We are God's vicegerents. We are meant to represent God and rule as God rules.

And how does God rule? With care and compassion, remembering his steadfast love, and listening for and hearing the cries of the suffering and oppressed. Such a conclusion, drawn from throughout the Old Testament, is reinforced by New Testament texts like Matthew 5:1-11 ("Blessed are those who mourn, for they will be comforted...Blessed are those who are persecuted for righteousness' sake, for theirs is the kingdom of heaven.") and Luke 4:16-20 ("The Spirit of the Lord is upon me, because he has anointed me to bring good news to the poor. He has sent me to proclaim release to the captives and recovery of sight to the blind, to let the oppressed go free, to proclaim the year of the Lord's favor.") So being an image-bearer of God for a Christian means imitating Christ—the Christ who shares in our mourning, washes our feet, and voluntarily takes up a cross for our sake.

Given that God is concerned for wildlands and wildlife, a Christian's love must also include them: humid wetlands and arid deserts; the stunning prairie orchid and the not-so-pretty Houston toad; soaring bald eagles and the burying beetle. God cares about *all* these places and creatures, and so should we, if we are to reflect the love of God. Our uniqueness as humans does not exempt us from extending care, but rather summons us faithfully to exercise care. In other words, human uniqueness is not a license for exploitation but a call to service. In a properly expansive Christian vision, worrying about wolves and warblers is part and parcel of the good news that nothing is beyond God's wide redemptive embrace.

Finally, Christians ought to care about the earth because it is a fitting response of gratitude. Care for our planet and its inhabitants is an appropriate way of saying "thanks" for God's bountiful and gracious provisions. As the old hymn, "For the Beauty of the Earth," puts it, "Lord of all to thee we raise, this our hymn of grateful praise." Gratitude is the grammar of a grace that fosters respectful care for God's creatures. We care for God's creatures because it is the appropriate response to God's providential care for us.

This, in my view, is the most compelling reason to care for the earth. The practice of gift-giving suggests that the experience of receiving provisions graciously offered evokes a response of gratitude and care. In other words, when given a gift, especially a valuable gift that meets basic needs, the appropriate response is gratitude to the giver and care for the gift. Grace begets gratitude and gratitude care. Thanksgiving not obligation drives the Christian moral life.

Despite all these reasons for being compassionate stewards of the earth, Christians have not had a sterling track record when it comes to treating the planet well. To be fair, neither have other cultures done well in this regard. Plato describes deforestation in ancient Greece. St. Augustine laments desertification in fourth century North Africa. The great Mayan cultures of Meso-America collapsed around the year 800 A.D. due to deforestation and soil erosion. Ecological degradation is no respecter of religions. It predates Christianity and can be found in places where Christianity has not been an influence.

Nor can Christianity be held solely responsible for the modern era's hegemony over nature through the tools of science and technology. While science and technology have certainly played a role in contributing to the current ecological crisis, other factors, especially economic ones, are equally if not more important. For example, the economist Bob Goudzwaard argues that modern capitalism has been a significant contributor to environmental degradation, and the historian Donald Worster makes a convincing case that materialism—both economic and scientific—is the principal cause of our ecological malaise.

Consequently, there is cause to doubt Lynn White's premise that Christianity gave science and technology the go-ahead to dismember nature. As Christian ethicist James Nash cautions, "The single cause theory for the emergence of our ecological crisis is pathetically simplistic. [C]omplainants...have accused Christianity of being the parent of ecologically debilitating forms of industrialization, commercialism, and technology. However, in historical reality, many complex and interwoven causes were involved—and Christian thought was probably not the most prominent one."[11]

Yet, even with all these caveats, it is important for Christians to acknowledge the need for confession about the excesses of their religion. Again, Nash offers us worthwhile commentary: "We cannot so easily distinguish between the faith and the faithful. The fact is that Christianity—*as interpreted and affirmed* by billions of its adherents over the centuries and in official doctrines and theological exegeses—has been ecologically tainted...The bottom line is that Christianity itself cannot escape an indictment for ecological negligence and abuse."[12] Nash goes on to say that Christianity has "done too little to discourage and too much to encourage the exploitation of nature." Therefore "ongoing repentance is warranted."[13]

Wendell Berry minces no words in agreeing with Nash. "Christian organizations, to this day, remain largely indifferent to the rape and plunder of the world and its traditional cultures. It is hardly too much to say that most Christian organizations are as happily indifferent to the ecological, cultural, and religious implications of industrial economies as are most industrial organizations."[14] Like Nash, Berry rightly calls Christians to confession, and so do I: ongoing repentance is warranted.

But that is not the end of the matter. Berry argues that however just the indictment of Christianity may be, "it does not come from an adequate understanding of the Bible and the cultural traditions that descend from the Bible." Those among us who are Christians must therefore learn to read the Bible anew precisely because our behavior is often out of line with the ecological vision and wisdom of the Bible itself—the very theme of this essay.

In sum, to allow the natural world to be ravaged, pillaged, and plundered is not only ethically wrong, it is religiously contemptuous. It is a sacrilege. To allow wildlands and wildlife to be lost is the ultimate irony for those who worship, in Berry's words again, "the wildest being in existence."[15]

Proper worship of such an undomesticatible God should find expression in grateful and joyful care for all that is holy—the silence of winter and the warmth of spring, aged oaks and scented pines, common loons and sandhill cranes, butterflies and little brown bats, rushing streams and oceans unfathomable as God's love. Christians have every good reason to preserve wilderness. The holiness of life on God's good earth evokes nothing less.

MOVERS AND SHAKERS IN THE
CHRISTIAN ENVIRONMENTAL MOVEMENT
Steven Bouma-Prediger

"I have never been able to entertain a God-idea," writes Joseph Sittler, "which was not integrally related to the fact of chipmunks, squirrels, hippopotamuses, galaxies, and light years."[1] Sittler died in 1987, but his spirit lives on in today's Christian environmental movement, a diverse group of thinkers and activists who bring a revised reading of the Bible—one of caring rather than dominion—to our relationship with nature.

One of the most savvy of these thinker-activists is Paul Gorman, the director of the National Religious Partnership on the Environment (NRPE), the country's largest interdenominational coalition of environmental organizations. It's comprised of the U.S. Catholic Conference, the National Council of Churches, the Evangelical Environmental Network (EEN), and the Coalition on the Environment and Jewish Life—virtually every major Christian and Jewish group in the United States. A sophisticated bridgebuilder and diplomat, Gorman has encouraged members of each group to explore the riches of their own tradition and reach consensus on certain substantive and strategic issues, without compromising their own principles.

Catholics, for example, draw extensively on a long tradition of church social teaching while Jews plumb the riches not only of the Hebrew Bible but also their oral tradition. Gorman is quick to point out that the coalition is driven by deeply religious motives: "We're not interested in being the shock troops for the Green Party. Care for creation has become a central element in religious life."

Even more important to note is the fact that several theologically conservative groups are also active in the Christian environmental movement. One of the most prominent is the Au Sable Institute of Environmental Studies, located in Michigan's northern lower peninsula. Offering a bevy of programs—retreats, conferences, college courses, and K-12 environmental education—the Institute has garnered a well-deserved reputation for excellence in its two decades of existence. Directed by the dean of the evangelical environmental movement, Calvin DeWitt, the Au Sable Institute weds rigorous environmental science with evangelical Christian theology.

DeWitt, an award-winning professor and widely read author from the University of Wisconsin, Madison, is also one of the founders of the Christian Environmental Council, a subset of leaders from the Evangelical Environmental Network who meet annually for reflection, fellowship, and action. That such a mustard-seed sized group can pack a political wallop was clearly evident in January 1996 when DeWitt led a contingent of Council folks to Washington, D.C. to lobby in support of the Endangered Species Act. Assuming that conservative Christian voices favored gutting the Act, members of Congress were caught up short when DeWitt's group captured media attention by showing up with an endangered Florida panther and a compelling Christian apologia for reauthorizing the Endangered Species Act. As Stan LeQuire, then director of the EEN, put it, supporting the Act was "a no-brainer" since it "resonated wonderfully with our biblical faith."

The EEN continues to address environmental issues like global climate change and protection of national forests, its mandate coming from a 1993 charter document entitled, "An Evangelical Declaration on the Care of Creation." Drafted by scholars and activists, this two-page brief sets forth a biblical vision of Christian earthkeeping. With now over three hundred signatories—including presidents of colleges and seminaries, chief executive officers of diverse Christian organizations, and a variety of scholars and activists—it is an indication of the depth and breadth of the movement.

One of the most creative and innovative groups is Target Earth, whose motto is "Serving the Earth, Serving the Poor." Led by Gordan Aeschliman, an energetic visionary who likes to howl at the moon, Target Earth hosts leadership training retreats for college students, helps organize lobby days in D.C., buys up tracts of wildlands in its version of the Nature Conservancy (called Eden Conservancy), and runs the Global Stewardship Study Program—a four-month-long college semester in the rainforest of Belize. With colorful folk like Peter Illyn, who barnstorms the Pacific Northwest with his llama Oochoo in search of converts to green Christianity, and Chris Eliasara, a Kiwi ex-pat and former rock climber who directs the Belize program, Target Earth is a kind of postmodern bricolage of environmentalists and social activists with the love of Jesus in their hearts.

This combination of ecological concern with social justice is by no means limited to Protestants. In 1999, the Roman Catholic bishops of the Pacific Northwest issued an eighteen-page pastoral letter which boldly set forth principles for sustaining the region's ecological and economic health. And Patriarch Bartholomew I, leader of the 300-million-strong Orthodox Church, made headlines around the United States in 1997 when he declared that "to commit a crime against the natural world is a sin." Whether Protestant, Catholic, or Orthodox, Christians are interested and involved in environmental work.

All is not sweetness and light among the Christian greens, however. The recently organized Interfaith Council on Environmental Stewardship (ICES) led by Robert Sirico of the conservative Acton Institute, finds many of the previously mentioned groups misguided or ill-informed. Issuing their own counter-document, the Cornwall Declaration on Environmental Stewardship, the ICES questions whether certain ecological degradations like global climate change or overpopulation are real, and claims that groups like the NRPE seek "to redefine traditional Judeo-Christian teachings on stewardship." But while the ICES worries that orthodox Christian faith will be compromised by environmental concern, many Christian environmentalists strenuously insist that care for the earth is integral to what it means to be a Christian. And given that the earth is groaning, to use the image of St. Paul in Romans 8, there is much to do.

But even if the planet weren't groaning in travail, many Christian environmentalists maintain that earth care is an essential feature of the gospel. Joseph Sittler captures this basic conviction nicely. "When we turn the attention of the church to a definition of the Christian relationship to the natural world," he writes, "we are not stepping away from grave and proper theological ideas; we are stepping right into the middle of them. There is a deeply rooted, genuinely Christian motivation for attention to God's creation, despite the fact that many church people consider ecology to be a secular concern. 'What does environmental preservation have to do with Jesus Christ and his church?' they ask. They could not be more shallow or more wrong."[2] Drawing on its roots in sacred scripture and faithfully bearing witness to God's vision of the flourishing of all creation, the Christian environmental movement is alive and well.

WORKING WITH FAITH-BASED ORGANIZATIONS
Suellen Lowry

Faith-based reasons for conservation have value in and of themselves. They also influence individuals and institutions to care for wild places, and have the potential to galvanize millions of people into lifestyle changes and political action for the environment. In the United States, 40–45 percent of the public consistently reports attendance at religious services in any given week.[1] As more of these people hear and respond to faith-based messages about caring for creation, the potential for significant societal change increases substantially.

For several years, a significant portion of my work has focused on promoting this sort of societal change by facilitating conversations between individuals in the religious community and lawmakers. Together, they have discussed faith-based reasons to protect wildlife and wild places.

I have been struck by the power of this work. In part, its power stems from the natural overlap between secular and spiritual conservation efforts, the importance of the religious voice to the general public and policy makers, the long history of social action in the religious community, and the synergy that results when various parts of the conservation community pool their talents.

It is my hope that people in the religious community and secular environmental groups will increasingly find ways to collaborate, and I offer the following observations to encourage such teamwork.

The Religious Community as a Natural Partner for the Protection of Wildlands

It is important to recognize that there is significant theological underpinning for the protection of wildlands. Religious and spiritual messages about the value of caring for creation have been a source of inspiration for centuries. Such messages are not specific to any faith and can be found in Christian, Jewish, and other texts. From St. Francis of Assisi and Hildegard von Bingen to John Muir and many modern environmental activists, individuals inspired in whole or in part by a spiritual calling have been concerned with conservation. In addition, there have been organized religious conservation groups in the U.S. since at least the 1970s.

There are many ways to become familiar with the variety of religious traditions that have set forth compelling principles for protecting the environment. Some of these writings can be found by accessing websites such as the following:

- Coalition on the Environment and Jewish Life (COEJL), *www.coejl.org*;
- Evangelical Environmental Network (EEN), *www.esa-online.org/ een*;

- Forum on Religion and Ecology, *http://environment.harvard.edu/religion/*;
- National Council of the Churches of Christ in the U.S.A. Eco-Justice Working Group, *www.webofcreation.org/ncc/Workgrp.html*;
- National Religious Partnership for the Environment (NRPE), *www.nrpe.org*;
- U.S. Catholic Conference Environmental Justice Program, *www.nccbuscc.org/sdwp/ejp/index.htm*;
- Web of Creation, *www.webofcreation.org*.

Each religious group also has issues specific to its particular tradition and locale. Therefore, it is helpful to read their publications, denominational or congregational web pages, and the Saturday religious page in city newspapers.

Important Considerations for Messengers

Often, people's willingness to hear a conservation message is based, at least in part, on who is delivering the message. In general, both policy makers and the public are inclined to listen to people speaking from a religious perspective. There are several types of these individuals:

- Those with affiliations to religious communities who are also active with secular environmental groups;
- Members of denominational and congregational social-justice and conservation committees;
- Students and professors at religious colleges and seminaries;
- Activists employed by national organizations like those listed above.

I have learned valuable lessons while working with people in the religious community. Here are some of them:

- I do not tell individuals with religious or spiritual messages what to say. Instead, I do my part to open the door so people from the religious community have an opportunity to share their beliefs and experience. Also, I only ask them to speak from their own areas of expertise.[2]

- In part because faith communities in the U.S. are very diverse, it's important to leave preconceived notions about individuals in the religious community at the door.
- In all instances, messengers' time constraints must be scrupulously considered.
- Whether reaching out to the public or policy makers, both lay and clerical messengers can be very powerful; and, when appropriate, it is also compelling to include other conservation messengers such as scientists.
- When the discussion is with an elected official, individuals whom the official represents—constituents or people with strong ties to constituents—are the most effective communicators.

For most meetings with policy makers, I encourage constituents from both the religious and scientific communities to meet together with their elected official, along with a local environmental policy expert. A typical meeting might begin with remarks about why we are called to care for God's creation, move to ecological information from the scientist, and conclude with information from the environmental policy expert. Each person speaks from his or her own area of expertise, and there is discussion throughout. Everyone has an important and complementary role to play.

Face-to-Face Dialogue

Whether reaching out to policy makers or the public, nothing can replace face-to-face communication and the opportunity the ensuing dialogue affords for building relationships. If we think about our own decisions to become involved in protecting wild places, we often find that it was personal experience that first convinced us to prioritize conservation efforts. For many, these personal experiences include being outside in wild country, but also for many, an important form of personal experience is face-to-face interaction in which someone's fervor for the wild is transmitted with the power of personal conviction. In addition, the "social capital" that results from face-to-face interactions can have far-reaching positive ramifications for conservation.[3]

With members of Congress, meetings in their district or state offices is a primary way to have a face-to-face dialogue. When engaging the public, face-to-face dialogue can occur in a number of settings, including worship services, congregational committee meetings, or Rotary and book club gatherings.

In short, dialogue that includes religious and spiritual viewpoints can powerfully promote the protection of wildlands. Work with faith-based messages and messengers can also be a very fulfilling personal journey.

∽

Editor's Note: For a more complete discussion of Suellen Lowry's work, see *Spirituality Outreach Guide* by Suellen Lowry and Rabbi Daniel Swartz, available from The Biodiversity Project, 214 N. Henry Street, Suite 203, Madison, WI 53703, 608/257-3513, *www.biodiversityproject.org*. Or contact Suellen Lowry at *suellen@northcoast.com*.

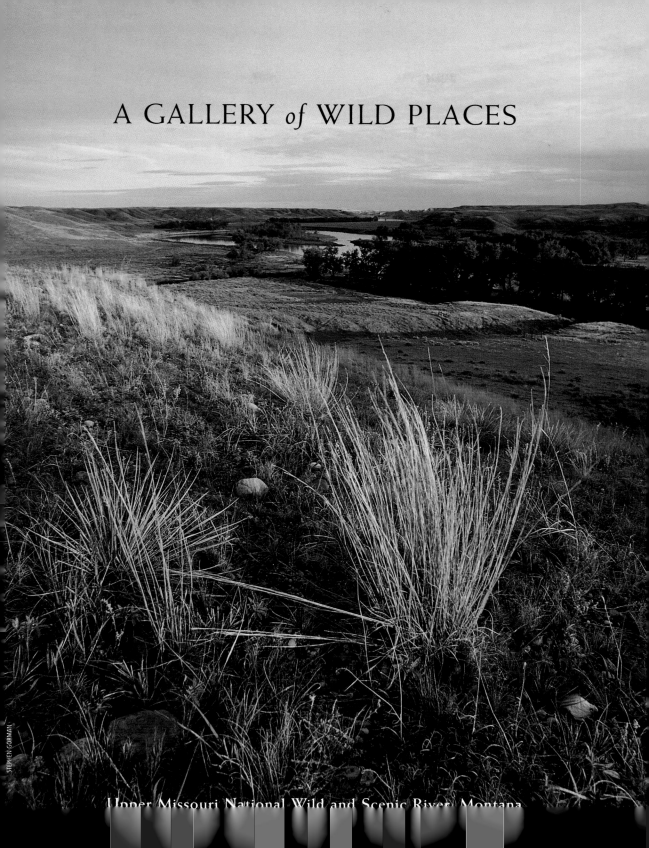

A GALLERY *of* WILD PLACES

Upper Missouri National Wild and Scenic River, Montana

∧ Pigeon Lake Wilderness
Adirondack State Park, New York

CARL E. HEILMAN II

JAMES W. KAY

STEWART AITCHISON

San Rafael Swell, Utah

< Parunuweap Canyon
Zion National Park,
Utah

Ten Thousand Islands
Everglades Wilderness,
Florida

Quetico-Superior Wilderness Area
Minnesota / Ontario

Dinosaur National Monument, Utah

Bridger Wilderness
Wind River Range, Wyoming

Mt. Hood Wilderness, Oregon

Lizard Head Wilderness
Colorado

Washakie Wilderness, Wyoming

Munds Mountain Wilderness, Arizona

Arctic National Wildlife Refuge, Alaska

PART III

WILDLIFE AND WILDLANDS

Wilderness needs no defense, only more defenders.

— EDWARD ABBEY

The Wild and Its New Enemies

Jack Turner

∾ It may seem peevish of me, but I detect a remarkable omission in the environmental literature these days. The emphasis is on social unrest, economic collapse, or the risks associated with further environmental degradation along lines that are well-known.[1] What is not mentioned, or at least not discussed in detail, are the issues that I believe will dominate environmental debate and wilderness policy in the near future: the replacement of wild beings and natural systems by genetically modified biological artifacts, cloned endangered species, and the potential introduction by nanotechnology of digital matter into wild ecosystems.[2]

The risks from introduction of what I think of as technologies of replacement are considerable, and I believe we are at present incapable of assessing them. Regrettably, the tendency of many wilderness lovers to dismiss these technologies as science fiction blinds them to a new reality. Science fiction it is not.[3] A recent report by the Rand Corporation for the National Intelligence Council concluded that by 2015 we will see a melding of biotechnology, nanotechnology, materials science, and information technology that will revolutionize technology itself. "Engineering of the environment will be unprecedented in its degree of intervention and control," say the authors.[4] It is precisely this intervention and control that destroys the wild by robbing it of its adaptivity and freedom.

By November 2000, a voluntary computer database called BioTrack maintained by the Organization of Economic Cooperation and Development (OECD) listed 10,313 field trials of genetically modified organisms in twenty-three countries.[5] In addition to genetically altered foods, such as soybeans, rice, and corn, that have received so much attention in the press, it turns out that many common organisms already exist in genetically engineered forms: coffee, eggplant, orange, apple, walnut, papaya, carnation, petunia, grass, salmon, halibut, cod, trout, mouse, cow, sheep, goat, poplar, pine, eucalyptus, Douglas fir, moth, mite, mosquito, worm, fungus, bacterium, and virus. Plus many, many more. The OECD claims that the number of genetically modified organisms in the BioTrack database is 1.6 percent of what we still so quaintly call the kingdoms of life.[6]

These organisms have been created by a process known as gene splicing, or genetic transfer. Genetic transfer takes a gene—the transgene—from one organism, splices it into a plasmid, replicates the plasmid in a bacterial culture, and splices one of the replicated genes from these new plasmids into another organism. The result is referred to as "transgenic," as in a transgenic tree, fish, or insect, etc.[7]

The first genetically altered insect to be loosed on the natural world is a pink bollworm moth enhanced with a jellyfish gene. It will be released into protected cages for field study in Arizona during the summer of 2001. Sterile but sexually active, this bollworm moth is designed to mate with wild relatives, thus gradually reducing the population of bollworms—an organism not favored by cotton farmers.[8]

Needless to say, introduction of similarly altered members of a species could reduce coyote populations, or the population of any other critter we happen to dislike. In the hands of those with malice toward the wild ecosystems, release of such alterations into the wild could be tragic.

Minor asides in the research literature suggest an even more intrusive and controlled biological future. At the Defense Department's Advanced Research Projects Agency's "Focus 2000" session, Dr. Robert Eisenberg, a professor at the Rush Medical Center in Chicago, gave a talk called "Bio Interfaces: Interfacing the Abiotic and Biotic World." A section on "biomachines" included discussion of "...insects made into smaller, lighter, faster, covert sensing and surveillance devices" and "engineered plants modified to be self-sustaining sensors of pathogens and toxins," presumably to detect an attack with genetically modified biological weapons.[9] The potential use of such weapons, especially by terrorists, is now a major concern of both the FBI and the CIA.[10]

Thus—in only several hundred years—humans have moved from suffering natural disasters like earthquakes, floods, and famine to fearing humanly manufactured disasters: climate change; chemical, nuclear, and germ warfare; technologically facilitated disease transmission such as the West Nile virus, which recently arrived in New York; and now a glut of artificial foods, beings, diseases, and weapons.

We need to distinguish this shift toward artificialization from the ubiquitous degradation of the natural world. The latter determines the health and integrity of our life support systems. The former is a matter of substitution: artifacts, simulations, and surrogates that replace the wild with objects generating entirely new levels of risk.

Something disturbing is at stake with all these replacements, something that strikes deeper into our souls than degraded ecosystems, the loss of species, or even new levels of risk brought on by the ever-accelerating advances of technology. It goes unnoticed because it cannot be seen with the eye, but it entails a vast disappearance with metaphysical, or, to be precise, ontological consequences: the material effect will hasten the end of evolution; the psychological effect will hasten the loss of the Other.[11]

An important part of understanding what we are results from defining ourselves in contrast to the Other, either as wild nature or the supernatural in one of its many cultural forms. With technologically advanced methods of artificialization, this particular sense of our self will begin to die. The world that was once both us and Other will become mostly us, a constructed world of physically and emotionally artificial humans, surrounded by their artifacts. The ancient authority of the wild world, the source of so much of our beauty and joy, will diminish.

In minor ways this sea change is upon us, and it leads me to a blasphemous edit of Thoreau's most famous adage. We can no longer claim that "in wildness is the preservation of the world." We must admit that "in wildness is the preservation of *this* world," for it is becoming obvious that more artificial worlds, perhaps even completely artificial worlds, are possible. With the construction of biological artifacts, we are creating and designing the elements of such a world as I write. I judge this set of developments to be more ominous than pollution, species loss, or climate change. It is a specter, and the leading edge of that specter is the process of gene splicing.

Some conservation groups, conservation biologists, and government bureaucracies are already considering, or actively pursuing, cloning and gene transfer, believing them necessary to achieve conservation goals. For instance, the Audubon Institute Center for Research of Endangered Species is interested in the use of cloning for endangered species preservation, and they are by no means alone.[12] John Varley, the director of the Yellowstone Center for Resources, has speculated on the possibility of using genetically modified organisms to remove exotic species from Yellowstone National Park: "…[T]here is no doubt in my mind that in the next ten to twenty years we will have genetically modified organisms that we can use as tools against non-native species."

Varley goes on to say, "But can you imagine the Environmental Impact Statement we'd have to go through? We all gnash our teeth because we have to use herbicides in a national park to control weeds like leafy spurge. What if someone came to us and said, 'I have this genetically modified beetle that can control leafy spurge.' Would we use it?"[13]

What about genetically modified rainbow trout immune to whirling disease, genetically modified elk and bison immune to brucellosis, or a genetically engineered beetle that would devour the billions of exotic New Zealand mud snails now infesting Yellowstone's waters? All are possible within the coming decades but not imminent. For those of us who love the wild, the creation of transgenic salmon and transgenic forests deserve our closer attention, since well-established research programs are about to unleash both of these manipulated forms of life on the planet.

TRANSGENIC SALMON

In the United States, Atlantic salmon is the most popular fish at the market. It is also the first genetically modified kind of fish (it is important not to say species) to be considered for human consumption by the U.S. Food and Drug Administration (FDA). If approved, this salmon, named the "AquAdvantage" salmon, will also be first kind of transgenic fish to reach our dinner plate. Its selection, its assessment, its approval, its future survival, even the amplification of its population, has nothing to do with evolution and everything to do with economics. According to the aquaculture industry, this transgenic salmon is just another farmed fish, a fish raised by the same methods, in the same kind of pens as currently farmed salmon, and exactly the kind of fish required to feed the world at a time of declining fish harvests.

This is one of the many instances of a biotechnology corporation's use of the biotechnology-to-feed-the-poor argument, claims that are inflated and fatuous. Most farmed fish, including salmon, eat fish meal made from small wild fish that are ground up. In the case of salmon, which consume "up to five pounds of wild fish for each pound of salmon produced," this amounts to at least 10 million pounds of wild fish killed each year. Hence, according to Daniel Pauly, a researcher at the University of British Columbia Fisheries Center, "The new trend in aquaculture is to drain the seas to feed the farms."[14]

Since eighty-two species of fish thought to be common are now considered endangered by the American Fisheries Society, there is reason for concern: traditional aquaculture depended on plants for fish food; the newer aquaculture industries tend to raise large carnivorous species such as salmon on ground up wild fish. This practice increases pressure on wild fisheries,[15] which many people worldwide—poor or otherwise—depend upon.

Oddly enough, the aquaculture industry, a 45-billion-dollar-a-year industry, at present opposes the commercialization of transgenic salmon because wholesale salmon prices have dropped, and the increased efficiency that would accrue from raising transgenic fish would probably lead to further decreases. They also oppose commercialization because they fear confusion on the part of consumers about the healthiness of transgenic fish—the same that clouded the introduction of transgenic crops.[16] Despite this opposition, 15 million genetically altered salmon eggs are stored at Aqua Bounty Farms in Canada, and, as of last year, 100,000 genetically altered AquAdvantage salmon swam in the company's tanks in New Brunswick and on Prince Edward Island and Newfoundland.[17] The company dismisses the industry's rejection of its product, claiming that everyone in the industry will eventually raise transgenic fish in order to remain competitive: if only one company begins raising AquAdvantage salmon, they say, others must follow—with all the freedom and scrutiny of dominos.

The Massachusetts-based parent company of Aqua Bounty, A/F Protein Inc., has applied to the FDA for permission to market AquAdvantage salmon to U.S. consumers. The decision in this test case is an important barometer of our ecological future. If A/F Protein's request is approved, AquAdvantage salmon could be on your dinner plate in 2002, the eggs would be sold to aquaculture companies worldwide, and approximately thirty-five more kinds of transgenic fish, crustaceans, and gastropods—including abalone, lobster, and shrimp—would begin their journey through the regulatory net to world markets.

Many scientists and environmental organizations believe this is an impending ecological disaster. Jean-Michel Cousteau, son of prominent ocean environmentalist Jacques Cousteau and an eloquent spokesman for marine conservation, puts the matter bluntly, "This second Green Revolution with species is sure to be the ultimate demise of what remains of our fragile ecosystems."[18]

Those who have followed the transgenic fish story are worried about two things: the effects of transgenic fish on their wild ancestral population, and their effects on other fish populations. In the former case there is reason to believe we may lose a wild species of salmon; in the latter we may intensify pressure on already declining fisheries.

AquAdvantage salmon carry a growth transgene from Coho salmon that increases its rate of growth; hence they are ready for market in eighteen months instead of the normal three years. Other genetic modifications can create larger than normal salmon. One Coho engineered in New Zealand weighed over five hundred pounds—a salmon the size of a grizzly bear.[19]

Other genetic alterations produce fish more tolerant of colder or warmer water than is normal for the wild species, or tolerant to lower levels of oxygen. Still others have improved disease resistance because of medicines in their food. Although these may seem like advantages, in each case unintended consequences for fragile aquatic ecosystems would be set in motion. Not only environmental groups but also professional organizations with impeccable scientific credentials have protested such genetic manipulation.

The Royal Society of Canada, for example, an independent group of scientists appointed by the Canadian government, says that if the Society's panel had its way "there would be a moratorium of GM [genetically modified] fish grown in farms on Canada's coasts."[20] The American Society of Ichthyologists and Herpetologists agrees, saying it "strongly favors a moratorium on creation or marketing of transgenic salmonids until it is firmly established that such fish will not gain access to natural waters, by accident or *intent* [emphasis added]."[21]

Such a moratorium would require that growing tanks be located inland and guarded against theft and the malicious introduction of altered fish to the ocean. An uneconomical prospect at best, it is patently absurd given the variety of countries farming transgenic fish—Cuba, Chile, Taiwan, China, Thailand. But the American Society of Ichthyologists and Herpetologists' statement underscores the salient point, and it has nothing to do with human reactions to transgenic salmon as food: the over-riding concern is that transgenic salmon will escape.

Given the record, scientists have good reason to be concerned about escapees. In 1988, a million of them escaped in Norway. In 1999, 300,000 escaped a farm in Washington. And an estimated 280,000 escaped from farms in British Columbia between 1991 to 1999.[22] Despite corporate predictions that the escapees would die in wild habitat, escapees from Washington and British Columbia have already been found as far north as the Bering Sea, and adult escapees are now caught by fishermen in the Gulf of Alaska.[23]

In December 2000, around 100,000 farmed (but not transgenic) salmon escaped into Machias Bay off the Maine coast only a month after the U.S. government declared the wild salmon in eight Maine rivers to be endangered due to a 65 percent reduction in North America's only wild population of wild Atlantic salmon. In January 2001, another 3,000–5,000 escaped in the same area. According to the Atlantic Salmon Federation, the number of farmed escapees is one thousand times greater than the number of documented wild adults.[24] The escapees mate with wild fish, and their offspring are weaker than the wild stock. And, of course, they compete with wild salmon for increasingly scarce resources. If these escapees were transgenic salmon, we would be faced with a potential ecological tragedy, unfolding along the following lines: Since transgenic male salmon are larger than wild salmon, they have a sexual advantage—as a rule, females prefer larger mates. The genetically altered offspring that are produced are less capable of enduring the rigors of the wild and succumb more readily than their wild ancestors. Given a long enough period of time, these generations of weak transgenic offspring will incrementally diminish the hardiness of the wild population. Scientists at Purdue University have developed a computer model that shows that as few as sixty transgenic fish in a population of sixty thousand wild fish will push the wild species to extinction in forty generations.[25]

This is the so-called Trojan gene effect, a demonstration of how one kind of biological artifact can replace a wild species. There would still be oodles of transgenic salmon left, in farms and in the open sea, but the world of wild salmon would be gone.

A/F Protein insists this won't happen because AquAdvantage salmon are sterile, but independent scientists say that 100 percent sterility cannot be achieved.[26] And if the fish are sterile, and do escape, then wild populations will also suffer. Sterile fish mating with wild fish will have the same effect on the wild salmon population as the sterile pink bollworm moth mating with its wild kin: it will reduce the wild salmon population. Indeed, reduction of an insect population that we perceive as a pest is precisely the point of genetically induced sterility. But in the case of wild salmon, transgenic salmon escapees will decimate a wild population we all value.

Furthermore, the mixing, loss, and confusion surrounding genetically modified crops suggests no reason to expect that the purveyors of transgenic fish could keep transgenic and wild populations separate. Their assurances that they could echo those of other high-tech companies—the nuclear and chemical industries come to mind—with this essential difference: for escaped transgenic salmon there is no half-life to the problem and no product recall. Like the loss of a species, genetic pollution is forever.

Medicines for farmed fish generate a similar dilemma. If farmed fish are given medicine and escape, they enjoy an advantage over wild populations which lack such protection. If they are not given medicine and escape, they may well introduce an exotic disease into the wild population. In either case, the ancestral population suffers.

How will escapees affect the rest of an aquatic ecosystem? Because of their genetically increased growth rate, transgenic escapees would eat more than wild salmon and thus place an additional burden on the ecosystem. And since many farmed fish are predators, they eat other fish; if they were also altered to tolerate a colder or warmer than normal water temperature, they could move into new ecosystems and become exotic invaders. This eventuality has already proved disastrous: after Nile perch were introduced to Lake Victoria, 50 percent of the fish species there simply disappeared.[27]

Multiply the varieties of transgenic fish, crustaceans, and gastropods waiting for the market by the number of all aquaculture farms in countries pursuing aquaculture—farms bordering virtually all oceans—and you have a big number. The coast of Maine alone harbors forty-five Atlantic salmon farms. Then, in your imagination, factor in the effects of all these possible escapees on other wild species. I, for one, begin to appreciate the magnitude and complexity of the problem. When I do this, I also feel the weight of Cousteau's concern: the ultimate demise of what remains of our fragile ecosystems.

FRANKENFORESTS

Oregon State University is the most important university in the world for transgenic tree research, the home of the Tree Genetic Engineering Research Cooperative, a $50 million powerhouse that intends to put transgenic trees into commercial production. In March 2001, a group of eco-terrorists, alleging themselves to be former students and alumni, attacked three sites where forestry professor Steve Strauss was conducting field trials of transgenic trees, primarily poplars that contained the pesticide Bt in their cells and were "Roundup ready"—immune to Monsanto's Roundup herbicide. Twelve hundred trees were destroyed.[28]

This act focused public attention on the fact that the debate surrounding genetically altered corn, soybeans, and potatoes had expanded to forests. Approximately 135 kinds of transgenic trees await approval in the United States, including fruits such as apple and grapefruit and forest varieties such as cottonwood, pine, fir, and eucalyptus.[29] Scientists are field-testing these trees, and various consortiums of universities, corporations, and governments hope to establish commercial plantations within five years.

Each of these new kinds of trees poses risks, both to its ancestral kin and risks to forest ecosystems. Although these risks are technically distinct from those surrounding transgenic fish, the deeper issues are the same: the replacement of wild species by manufactured simulacrums; the threat of losing valuable wild populations; and natural selection being driven by economic factors, not environmental ones. In sum, the decline of the wild.

However, I believe the prospect of transgenic forests to be of special importance to lovers of American wilderness. In his "Fact-Book," Thoreau noted that "wild" is the past participle of "to will"—self-willed—and recent scholarship suggests that wilderness is best thought of as self-willed land.[30] This explains what we mean when we speak of the wild as untamed, untrammeled, uncontrolled—it gets along just fine by itself without our help or direction. It is also likely that deep in our linguistic past, "wild" was related to the German "wald" or forest.[31] Given that Europe and North America were once covered by vast forests, I think our core idea of wilderness originated in primeval forest ecosystems with their fecundity, self-perpetuating organization, chaotic complexity, and zillions of wild critters. The core idea of wilderness in the Bible or in a seafaring culture might be different, but our wilderness stories are often of the forest.

Ninety-five percent of America's old growth forests have been destroyed, 75 percent of the world's old growth forests have also been decimated, and the lumber companies are doing their best to cut down the rest to satisfy the 400-million-dollar-a-year global wood products industry.[32] What do we need in place of this old growth to satisfy our ongoing appetite for wood products? The answer from some very important institutions with considerable authority and economic clout is simple: vast plantations of cloned transgenic trees.

The lumber corporations behind this program are the ones you would expect—Alberta Pacific, Boise Cascade, Fort James, Georgia Pacific, International Paper, Potlatch, Westvaco, MacMillan Bloedel, Union Camp, and Weyerhaeuser. Other backers of transgenic trees are less obvious—for instance, Monsanto, Mycogen, Shell, and Toyota, not to mention numerous universities, the U.S. Department of Agriculture, the U.S. Department of Energy, and the U.S. Environmental Protection Agency. And that is only in the United States; sixteen other countries are also conducting field tests of transgenic trees.[33]

Monocultural tree plantations are the opposite of a wild forest. They were developed about two hundred years ago, in Germany, and conceptualized along the lines of a factory or industrialized agriculture. Owners wanted a simplified, standardized, predictable forest that would maximize the return on a measurable commodity—a forest crop of board feet or cords of firewood. Gifford Pinchot, the first chief of the U.S. Forest Service, made the analogy with industrial agriculture explicit: "Trees may be grown as a crop just as corn my be grown as a crop...the farmer gets crop after crop of corn...the forester gets crop after crop of trees."[34]

Everything inimical to maximizing production of this commodity was deemed a weed or a pest and either eliminated or deprived of sustenance: grasses, flowers, fungi, lichens, ferns, mosses, shrubs, vines, reptiles, birds, and insects. Other human uses of forests such as hunting, gathering, fishing, trapping, worshiping, seeking refuge, and practicing rituals were banned on these plantations.[35]

At first these intensely managed factory forests were profitable; it took a hundred years for the owners to realize that something was dreadfully wrong. Production waned for reasons that are now obvious: nutrient cycles were disrupted; nitrogen fixation dependent on lichens stopped; and the soil, which is primarily insect poop and the remains of microorganisms, became less fertile. As James C. Scott notes, "Only an elaborate treatise in ecology could do justice to the subject of what went wrong..."[36] And it is important to add that such a treatise still can't be written, for we do not yet know enough to understand a single forest ecosystem.[37]

Nonetheless, the managed forest plantation remains the model for both private corporations and public institutions, and transgenic forest plantations will build on that tradition. From the companies' point of view, a cloned transgenic tree is just another advance in standardization, an extension of what is already normal. Unfortunately, what can "go wrong" is still with us, and transgenic forests will only exacerbate it, though again it may take us a hundred years to notice something is wrong.

At present there are a variety of major areas of research in transgenic forestry. Each has its own potential problems, not all of which concern the environment; together they suggest the breadth of what will be a revolution in what we view as a tree: plum trees resistant to disease; apples that will not turn brown when exposed to oxygen; and "cherries in a variety of fashionable new colors" (fifty field trials of these brave new orchards have already been approved).[38] The problems here are familiar from our brief history with transgenic crops: built-in "enhancements" to increase a crop's market value can cause genetic pollution of non-transgenic fruit and nut trees and worries about possible adverse effects on the human food supply.

Another important area of research seeks to produce a tree with reduced lignin. Lignin is present in the cell wall of trees and it must be removed to convert the wood to pulp suitable for the production of paper. Unfortunately, removing lignin from trees creates unpleasant smoke stack emissions and puts toxic sludge into rivers. This makes genetically modified trees with reduced lignin look like progress. But lignin also has positive properties: much of a tree's strength is due to lignin. If transgenic trees with reduced lignin pollinated wild forests, they would reduce the strength of wild trees, and natural events such as windstorms—normally benign—could create havoc in native forests.

Trees are also being developed that absorb dioxin and PCBs, and trees that are drought resistant are being created by countries characterized by desert ecosystems. Israel, ever short of water, has created transgenic poplars that will green the Negev desert, for example. But desert ecosystems are also fragile and it is hard to imagine that transgenic tree plantations large enough to make a difference would not have profound effects on local ecosystems.

Despite the variety of these programs, and their potential for harm, two research regimes stand out as especially pernicious. The first concerns transgenic forests that contain a pesticide and are also "Roundup ready," i.e., tolerant of Monsanto's popular herbicide. These are the trees most likely to replace formerly natural forests—pines, poplars, firs, cottonwoods, cedars—species that are already economically important for the production of lumber, wood pulp, or chipboard. Looked at from an economic perspective, vast plantations of transgenic trees are simply more efficient at producing wood than natural forests.

Second, and perhaps more important, intensive research is underway to produce transgenic trees that absorb carbon dioxide more efficiently than natural trees. A new parlor game exists that speculates on how much new forest we might need to solve the greenhouse gas problem. Norman Myers guesses that a eucalyptus or pine plantation the size of Zaire might do it; a scientist at Oak Ridge National Laboratory suggests a tree farm the size of Australia; another think-tank suggests one half the size of the United States.[39]

Whoever is correct, they are obviously talking plantations of unprecedented size, and we are well on our way to planting them. Heavy producers of carbon dioxide have already contracted with southern nations to grow forests in exchange for what they hope will be "pollution credits." Such credit will allow them to pollute with impunity because of forests elsewhere on the globe.

Four million hectares of carbon gulping plantations already exist;[40] the

Netherlands has contracted with Malaysia and Ecuador; the United States with Costa Rica; England with Uganda; Canada with Queensland, Australia—the list is long. The logic is clear: first you get the profits from cutting down the forest, then you replace it with a forest that is even more profitable because it is more efficient—faster growing, pest resistant, and herbicide tolerant—then you sell pollution credits to help a corporation avoid the cost required to reduce its contribution to global warming.

The Bush administration's recent withdrawal from the Kyoto Protocol, which was designed to reduce pollution along with global warming, may accelerate the quest for carbon sequestration as many industrial countries discover that the creation of vast carbon dioxide gobbling forests, especially ones located far from home, are politically preferable to economically unwelcome reductions of greenhouse gases. Thus a large and growing international market may soon exist for a transgenic tree that is more efficient at reducing carbon dioxide than current tree crops.[41]

Toyota Motor Corporation is developing just such a tree. It has its own biotechnology program and aims to clone its winner, over and over, to "have whole forests of these extra-efficient 'super trees,' to help purify the atmosphere."[42]

Given all the wonderful things that transgenic forests can do for us, why would anyone object to their development? The answer has many layers.

The first difficulty is that the agency charged with permitting field trials, the U.S. Department of Agriculture's Animal and Plant Health Inspection Service (APHIS), allows applicants to put "CBI"—"confidential business information"—in the space describing the transgene being tested and its origin.[43] This renders it impossible for independent scientists to analyze the risks of the experiment, much less the effects of the resulting tree on an ecosystem.

As with transgenic fish, the troublesome aspects of transgenic trees can be divided into those that affect their ancestral kin and those that affect the rest of the ecosystem. Like fish geneticists, tree geneticists must deal with the issue of sterility. If the new transgenic tree is not sterile, then it can contaminate its kin, domestic or wild, by crosspollination. Looking at all the evidence, a professor of agricultural engineering at Iowa State University recently concluded this is what likely happened with StarLink corn—a genetically modified crop that was approved only for animal consumption but found its way—either through crosspollination or a human sorting error—into food destined for humans.[44]

Unfortunately, trees are even more genetically compatible with their wild kin than intensively hybridized food crops, hence crosspollination will be more problematic.[45] Tree pollen has been found on the treeless Shetland Islands 155 miles from its source; and pine pollen in India was blown 375 miles from its point of origin.[46]

Furthermore, common pollinators—flies, butterflies, ants, beetles, aphids, and bees—do not respect property boundaries. Crosspollination could even occur from waterborne leaves and twigs, or from bacteria and fungi,[47] as well as from the 25 million airborne insects that occupy the airspace above a single square mile of Earth's surface.[48]

The idea that we might regulate crosspollination from plantation to native forest is nonsense. Many transgenic plantations in the United States would be in close proximity to wilderness areas, and as far as I can tell, no wilderness area outside Alaska is more than 375 miles from where a transgenic plantation might be planted. Many non-sterile transgenic trees would probably pollinate their wild kin.

Then, too, since transgenic trees would grow for years before being harvested, there would be plenty of time for them to crosspollinate. And if they did crosspollinate, the receptor tree could live for centuries, colonizing a remote corner of wilderness with no one knowing about it—much less knowing what to do about it. Again, there will be no half-life for the problems, no product recalls. Genetic pollution into wild ecosystems is irreversible and every instance of genetic pollution will further diminish the wild.

Companies will patent their successful transgenic trees, so they won't "lose" the genes to, say, a competing company; they thus have a compelling reason to create sterile trees. Unfortunately, keeping transgenic trees sterile is even more unlikely than fish sterility. Generations of trees can change the order and expression of the introduced gene, especially if confronted with environmental changes and stress from disease.[49] Hence there is no reason to believe wild ecosystems will not be vulnerable.

Even if these vast plantations are planted with sterile transgenic trees, how will they affect neighboring forest ecosystems and other constituent environments—geology, aquifers, wind, and climate? How will the virtual elimination of insects, flowers, and seeds change, say, bird populations and their migration routes?

And what about issues of scale? Are we to extrapolate from what we learn about a protected plot of an acre or less to the potential consequences millions of acres of transgenic plantations worldwide? To do so would be a rather pathetic inductive leap of faith. No one knows what unintended consequences might arise, but some effects of replacing natural forests with transgenic surrogates are apparent even to a layperson. Give the issue a few minutes, and the questions multiply like rabbits.

Faster growing trees will require more water, a substance becoming as precious as gold, not only in the American West, but worldwide. From a lumber company's point of view it may make sense to spend billions of dollars (especially if much of it is public money) to create trees that guzzle more water than their natural counterparts. But nearly everyone else that lives in the American West has a stake in the issue—lovers of wild rivers, anglers, ranchers, farmers, energy consumers—and they deserve to have their interests represented *before* such plantations become a *fait accompli*.

Transgenic trees will also require large amounts of fertilizer because the complex food webs of soil microorganisms will be reduced or eliminated by intensive use of herbicides. Since we haven't even discovered, named, or counted these beings, shouldn't we refrain from eliminating them over vast areas of the globe? Whatever happened to Aldo Leopold's idea of saving all the cogs and wheels?

Insects targeted by a genetically introduced pesticide will also likely interfere with insect population dynamics, with unknown consequences for their wild kin and other species higher up the food chain. How will a tree that kills insects on contact affect wasp, lacewing, and our beloved ladybug populations?

Given our ignorance and this morass of possibilities, it strains credulity to think that anyone can conduct reliable "risk assessments" of transgenic trees. But assessments must be done—the public demands it. And assessments will be done. By whom? The Environmental Protection Agency had to choose some group to assess the risks of transgenic forestry. And it did: the Tree Genetic Engineering Research Cooperative at Oregon State University, the very folks interested in commercializing transgenic trees.

CONCLUSION

All this is so sad, given the alternatives. Before the Columbia River was dammed, between 10 and 16 million salmon swam up its channel each year, and that is only one salmon river out of thousands.[50] If we were a more ecologically friendly society, we wouldn't have destroyed that fishery, and we would not now need aquaculture and transgenic salmon. Similarly, scientists at the Andrews Experimental Forest in Oregon wouldn't be placed in the defensive position of claiming that ecologically friendly forestry methods—instead of transgenic ones—can provide us with both wild forest and lumber.[51]

But once a society starts down the road of technological fixes, it becomes addicted, and what ensues is a cascade of fixes, hundreds of them, each creating the need for more fixes.[52] That's where we are now—somewhere near the falls and ready to go over the edge, with many of us ready to alter the entire earth that has sustained us. First we had Frankenfood, now we face the prospect of Frankenfish and Frankenforests. Where will it end?

At the same time that the wild is confronted with such disasters, what was once at least a somewhat united environmental movement continues to bifurcate along lines of individual obsessions. Many people, I fear, will be seduced by the carrot of a techno-fix for their favorite environmental malady.

Some who are obsessed with climate change will accept transgenic carbon dioxide-sucking tree plantations, regardless of the consequences. Some who are obsessed with social justice will embrace transgenic fish to feed the starving poor and transgenic forests to provide them with the warmth of fires and the succor of a home. Some who are obsessed with preserving biodiversity believe vast artificial tree plantations will reduce pressure on our remaining old growth forests. Some, many now, are obsessed with fun and could care less if their outdoor playground is wild or as artificial as a beer can.

One cannot overstate how destructive these bifurcations have become. We are losing the soul of our planet. The words that rolled so easily from the lips of early conservationists, *wildness* and *wilderness*, now refer to concepts of secondary, even tertiary, importance to most people. We are abandoning the wild at the very moment it most needs our ministrations.

If these bifurcations endure and one is forced to choose, then I choose to stand with those who believe in the primacy of the wild. I think of us as Thoreauvians. Civilization has plenty of champions, as Thoreau pointed out long ago, but the new genetic technologies require us, more than at any time in history, to champion wild nature. May we turn aside from the ever troubled spectacles of human invention and economic woe, to cling fast to the wild world, humbled as it is, and defend it to the end.

Should Wilderness Be Managed?

Michael E. Soulé

∾ PROEM

Wilderness in the traditional American sense (and in the Wilderness Act of 1964) is often defined as self-willed, autonomous, untrammeled, free, independent, or undeveloped implying that the essential quality of wilderness is its separateness from human control or manipulation, if not from sojourning backpackers, hunters, and munching livestock. Nevertheless, wilderness, along with the entire surface of this planet, is afflicted by civilization and its products, and is less than perfectly self-willed. This awareness of the universality of human-caused wounds encourages some humanitarian critics to argue that all land, excepting perhaps Antarctica, is a defiled human construction, and by default, is an appropriate venue for human economic expansion and "gardenification." It might seem academic, therefore, to argue about whether wilderness ought to be managed.

Indeed it is true that past and present activities such as human occupation (aboriginal or modern), burning, grazing, logging, mining, and planting do compromise the absolute independence of Nature and wilderness.[1] It is also regrettable that management actions now being undertaken to correct the changes wrought by past defilements are becoming a cause of bitterness among Nature's defenders—conservationists. For example, Jack Turner criticizes The Wildlands Project for planning a network of larger and better connected wilderness areas in North America on the grounds that the system is artificially created and visible from space: "If these designs were done as The Wildlands Project envisions...[it] would be the largest artificial structure on the planet. Sorry dude, but I don't want North America to be a planned community based on untested science."[2]

Perhaps it is the quest for perfect freedom that leads many people to wish for perfect wildness in Nature. But even as we are inspired by this ideal, the "real work"—as Gary Snyder calls it—is our communal struggle to protect the beauty and integrity of nature, a project that is necessarily sullied by the expediencies and compromises of politics. The ideologically diverse participants in this work are often tempted to circle the wagons and shoot inward—a habit of conservationists. Instead, we might consider riding a mile in each other's wagons.

THE FALL: A WOUNDED NATURE, A PRAYER FOR HEALING

Some of the wounds on Nature's body are old and scarcely detectable without paleo-forensic research. The most profound of these is the absence of the megafauna—dozens of species of large animals, including mammoths, mastodons, giant ground sloths, native horses, camels, American lions, and cheetahs. They disappeared about eleven thousand years ago, soon after the arrival to North America of sophisticated hunters from Asia. The megafaunal extirpation might be called "ecological decapitation," since Earth's major ecological players are likely to have governed the structure and diversity of their ecosystems, just as large mammals still do in parts of Africa where they persist. This means that most places that are called "wild" or wilderness today are ecologically truncated, and much less wild than they were in the past.[3]

It is probably no coincidence that the majority of the species of large mammals that survived the die-off eleven thousand years ago had co-evolved with spear-throwing humans in Asia during the last glacial period. These circumspect survivors include species like elk, moose, wolves, grizzly bears, and beaver. Even these, however, were exterminated from large parts of their former ranges in Europe and North America during the last few centuries as technological activity spread over the continents, initiating the modern era of trade in wildlife and economic globalization.

We see the consequences of the latest phase of the removal of large carnivores in the superabundance of white-tailed deer in the American Midwest and East, elk in protected places like Yellowstone and Rocky Mountain National Parks, and in the mushrooming number of smaller, bird-eating carnivores like raccoons. These species flourish in the absence of large carnivores, particularly the wolf. The ecological effect of dense populations of white-tailed deer, for example, is the over-consumption of understory plants, including the seeds and seedlings of forest trees such as aspen, oak, and hickory. This has led to dramatic changes in forest structure and composition in many regions.

The more that one knows about the ecological and land-use history of a region, the more one is saddened by missing pieces, including predators, amphibians, fish, and the more one's joy is checked by invasive species and by the virtual absence of reproduction in some forest trees because of too many lounging herbivores.

In addition to missing species, there are the more obvious wounds such as pollution and the over-growth of algae in freshwaters due to nitrogen-rich run-off from fertilized fields, lawns, and golf courses. Moreover, nest-robbing species such as ravens and magpies have become superabundant because they are subsidized by farm wastes, pet food, compost piles, and roadkills. Other superabundant birds (starlings, rock doves, house sparrows), plants (tamarisk, purple loosestrife, kudzu, cheatgrass), molluscs (zebra mussel), fish, and rodents have been brought to North America without their restraining pathogens, parasitoids, herbivores, and competitors. Introduced organisms, including diseases and pests, are changing the composition and structure of ecosystems globally.

For example, the American Southwest is replete with the scar tissue of ecological abuse. To someone sensitized to the signs of past overgrazing, for example, it is impossible to ignore the gully and sheet erosion, the absence of perennial grasses and living soil crusts, and the resulting rise to dominance of plants such as creosote bush and prickly pear cactus. Ironically, though, the stark badlands created by past abuses can create landscapes that appeal to artists and tourists, and may even enhance their experience of the "wild." A good example is the cattle-burned landscape between Santa Fe and Abiquiu in New Mexico made famous by Georgia O'Keefe. Ignorance is indeed bliss and it is often economically beneficial to the tourism industry in disturbed, biologically artificial destinations like the Hawaiian Islands.

A place that appears wild and natural to the naïve trekker, climber, mountain biker, skier, outfitter, or tourist may stink of sickness and signs of slow degradation to the ecologist. Compassionate naturalists, therefore, are motivated to fix a wounded ecosystem in order to maintain or restore its biological diversity and resilience. To such a person, the wildness of a place is compromised in proportion to its ecological degradation; a place is even more upsetting if even greater degradation in the future is probable. The proliferation of roads—perhaps the most potent cause of denaturation—is the worst wound of all. Road impacts may spread and accumulate over time, becoming open wounds that never heal.[4]

Non-ecologists, including wilderness adventurers, may experience insults in the wild as well, but their concept of the wild and of wounds to the wild often differ from those mentioned above. Solitude—the relative absence of other humans beings, including aircraft—is the most important defining characteristic of the wild for many non-ecologists. To them, a peopled place is a wounded place. The easiest material wound to detect is the jetsam of consumer economies—trash. Our subjective transaction with trash taints the wild by coloring our experience. Usually, though, trash does little harm to terrestrial ecosystems. Even abandoned cars and plastic bags and containers vanish in time. Other kinds of visual pollution, such as microwave relay towers on mountain tops offend the seeker of pristine Nature more than they harm ecological processes and biological diversity.

In other words, the concept of a wound can be highly subjective. The current mass extinction and universal ecological degradation is the private agony of naturalists, ecologists, and other lovers of wild things and places. To cope and carry on, they beat their breasts, rage at other conservationists, armor their hearts, and defend Nature against the well-meaning critiques of humanitarians who justify the biocide of the few remaining wild places as a necessary survival mechanism for oppressed, unprivileged classes of *Homo sapiens*.

Some humanitarians may have little compassion for the pain felt by naturalists, just as some biologists may not believe that social justice issues should always trump the needs of non-humans. The notion that compassion can be extended to distressed non-human beings and inanimate natural objects, and that such extra-human compassion is just as virtuous and heartfelt as compassion extended to suffering people, is a difficult one to accept for professional humanitarians who speak for less powerful groups in the Third and vanishing Fourth Worlds. These critics often condemn, using terms like "misanthropic" or "racist," the idea of untrammeled wild places free of human economic exploitation.

Another group—we'll call them managers for now—may have very different criteria for the "woundedness" of nature. Utilitarian conservationists may react with disgust to a landscape that is underutilized or not optimally productive of some commodity. To pastoralists, a grassland that isn't grazed is offensive. And I have known foresters who are offended by a stand of ancient redwood trees because they can imagine a "healthier" alternative—a even-aged stand of young, rapidly growing redwoods. To a manager of a national park who believes deeply in the benefits to society of wildlands recreation, the absence of roads, formal camp sites, stores, and other tourist amenities, if not a wound, is tantamount to professional malpractice—a psychic wound.

THREE WAYS OF PERCEIVING THE WILD AND WILDERNESS

Clearly there exists a wide range of perspectives among those who mourn Nature's demise and the woundedness of those wild places that remain. Table 1 illustrates three perspectives on wilderness and its management. These archetypes are arbitrary; most of us probably fit at least two of them, depending on circumstances.[5]

Someone whose primary identification is with the first category—the managerial—loves the resources and services of Nature. Many managers enjoy hunting, fishing, and the physical and mental challenges of resource extraction and outdoor sports. Many professional wildlands managers derive satisfaction from providing citizens with the opportunities and pleasures of the great outdoors, and the protection and production of natural resources and commodities for future generations. Managers often consider their path to be a middle one that maximizes product and minimizes damage to the "resource." Management is not a necessary evil for them. Rather, it is a way of providing sustainable recreational and economic opportunities and creating harmonious, "win-win," scenarios through multiple use and compromise. If they worshipped a god, it might be Pales, the Roman goddess of shepherds and flocks; so I will call them Palesians.

Those who identify with the second archetype—the *ecological*—love the richness and interactions of living nature, its animals, plants, and food webs. Their god might be Artemis, the Greek goddess of wild animals, fertility, and the hunt. Artemisians are painfully aware that the natural world is severely wounded. They want to heal the ecological wounds, and are

TABLE 1
Three archetypic ways of perceiving and saving the wild (or nature)

	Managerial/Political	Ecological/Process	Heroic/Experiential[1]
Proponents	Agencies, managers, utilitarian conservationists	"New" conservationists[2]	Traditional conservationists, wilderness adventurers
Philosophy	Utilitarianism/Ethical, Hedonism[4]	Pragmatism,[3] Deep ecology (for some)	Positivism, Deconstructionism
Dead heroes	Roosevelt, Pinchot, Leopold, Marshall	Muir, Carson, Leopold, Shepard, Abbey	Muir, Marshall, Thoreau, Brower, Shepard, Abbey
Myths/Ideology	Resourcism, multiple use	Paganism, modern biology	Perfectability of society, bioregionalism
God	Pales	Artemis	Daphne
Goals for wilderness	Use, satisfactory resolution of conflicts among users	Protection of nature, experience of living nature and the wild	Experience of untrammeled space and the wild, solitude
Natural allies	Resource extractors, recreationists, hunters	Biologists, ecocentric recreationists, naturalists	Animal rights advocates, mountain climbers
Solution/ Methodology	Consensus, win-win	Planning, analytical, empirical, incremental improvement	Revolution, criticism, restricted usage
Implementation	Top-down, regulatory, user-driven	Activism, incentives, purchase, zoning of land uses, restoration	Hands-off, population reduction, radical change in lifestyle
Perceived obstacles	"Extremists" of all kinds and enemies	Wise use ideology, idealism, wheeled recreationists	Managers, scientists, most ecotourists
Perceived proximal causes of crisis	Proliferation of diverse interest groups	Population, technology, globalization of commerce	Population, scientism, centralization & management
Perceived ultimate causes of crisis	Human irrationality, orneriness	Human greed, arrogance[5]	Human greed, Edenic fall
The ends and means issue	The means are the ends	Ends justify less-than-perfect means	Ends never justify imperfect means

[1] Based largely on Jack Turner's book, *The Abstract Wild*.

[2] Includes those with a biological/ecological orientation toward nature and wilderness.

[3] "On pragmatic principles, we can not reject any hypothesis if concepts useful to life flow from it…" William James, *Pragmatism*, Lecture VIII.

[4] This refers to Jeremy Bentham's goal of welfare economics: achieving the greatest happiness for the greatest number.

[5] See D. Ehrenfeld, *The Arrogance of Humanism* (Oxford: Oxford University Press, 1978).

willing to tolerate a certain degree of intrusion and esthetic offensiveness in the interim, if that is essential for healing. Artemisians also have more concern for the protection of biodiversity (Creation) and natural processes than for the "experience" of wilderness in the short run. I am more in the camp of the Artemisians, in case you are curious about my bias. Many employees of management agencies are "closet Artemisians."

People in the third category—the *heroic*—love wild, free, untrammeled spaces, the grandeur of vast landscapes, the fierce simplicity of high mountain peaks, and the physical and emotional challenges of fast water or sheer rock. Their goddess might be Daphne, the Greek nymph who was pursued by Apollo, archetypally the god with an overview, the creator of order and harmony. Running from him (metaphorically fleeing from his organizing power) and certain that she would be caught and tamed, Daphne asked Gaea for help. Gaea, the goddess of Earth, turned Daphne into a laurel tree, forever wild, forever part of the forest. Daphnians might also be called "wilderness purists." Like all conservationists, they perceive the natural world as wounded, but in different ways than Artemisians and Palesians. Daphnians are offended by trash and other signs of civilization such as noisy, technologically equipped hikers; in addition, economic activity such as tree stumps and mountain-top communications facilities are disturbing to them.

Daphnians, at least the most pure, are characterized by their hostility to control and management of wilderness. They might engage in certain management activities such as picking up trash, but they would prefer that ecological wounds in designated wilderness be untreated, unmanaged, and unrestored, at least by mechanical or invasive means. Some Daphnians dislike research or management practices that interfere with the physical independence or dignity of animals by the use of visible, intrusive technologies such as radio collars, and tagging or marking.

I have emphasized the differences among the three categories of conservationists, but the agreements are at least as profound. The Artemisians and the Daphnians agree that human overpopulation, technology, and globalization are root problems. Those who subscribe to any of the three conservation archetypes may experience spiritual values in Nature, including the divinity of place, a transcendent sense of oneness, and feelings of humility, grace, or wholeness in wilderness. At a more mundane level, all varieties of wilderness supporters may find solace in the less numinous satisfactions of the wild, such as escape from pressures of city life, and the sights, smells, touch, and sounds of Nature. In short, all three love Nature.

So why are they often unable to resolve their differences and present a united front? A major roadblock on the path to common ground in the arena of management is the dilemma of "naturalness versus wildness"—or the vitality of the wild (naturalness) versus the appearance of the wild. Appearance is a social value—a cultural attitude that colors how we feel about a place.[6] On the other hand, vitality of naturalness is, in my view, an intrinsic value, neither swollen or shrunken by how wilderness visitors feel about a place.

Can we examine the issue from non-anthropocentric perspective? If, for example, all species could vote on the issue of vitality versus appearance, the result of the election would likely be an overwhelming plurality in favor of naturalness because plants and animals would, of Darwinian necessity, insist on persistence, a virtual synonym of vitality. Fortunately, some management interventions accomplish both objectives. For example, if society were to declare a policy of maximizing native biodiversity, it would require the repatriation of ecologically effective populations of large carnivores. Such rewilding requires larger, connected wilderness regions and ultimately minimizes the necessity of anthropogenic controls.[7]

MANAGEMENT: A MULTIPLICITY OF MEANINGS

Many debates about management exist because of diverse meanings and associations of the word, some of which elicit visceral reactions and some of which have nearly polar opposite connotations. For example, management can imply an attitude toward the wild of caring, constructive guidance, stewardship, custodianship, and require skills or qualities such as mediation, mentoring, organizational effectiveness, and efficiency. Correspondingly, many Artemisians and many managers themselves associate management with compensatory actions meant to repair the damage caused by past errors in land-use such as over-grazing, fire suppression, and the introduction of harmful species for economic reasons such as bufflegrass, salt cedar (tamarisk), and bullfrogs in the Southwest.[8] Naturally, to attack wilderness managers and conservation biologists as classes for such work creates befuddlement, if not instant enemies.[9]

On the other hand, the archetypic Daphnian may associate management with (1) technological domination and control, (2) the patronizing elitism of scientists, (3) and the flagrant manipulation of animals by special interests, as in the roundup and killing of bison on the borders of Yellowstone National Park. A recent exchange on a list-serve underlines why Daphnian anger is often directed at those in the management profession: "The Wildlife Society sanctions a day-long contest in New Mexico amongst students in which points are awarded for coyotes, foxes, and bobcats killed. The winner is the student who kills the most predators." No doubt these students and their mentors rationalize such contests with the argument that they rid nature of marauding predators that may cause economic harm to livestock owners and diminish the enjoyment of upland game hunters.

It is not surprising, therefore, that "management" can evoke images of surveillance, captivity, harassment, and constraints that restrict individual freedom. In fact, the abhorrence of control and constraint of human beings is often projected onto warm-blooded animals, conferring on them such human qualities as dignity and independence. For example, David Brower and other prominent conservationists argued that the removal of fertilized eggs from condor nests in the wild and the capture of all remaining wild California condors during 1987 for captive breeding and eventual repatriation (as is now happening) was a greater evil than allowing the species to disappear with dignity from nature.[10]

Similarly, Jack Turner has argued against the use of technological means to return wolves to the western United States because the capture, sedation, transport, temporary captivity, radio-collaring, and release into a strange part of the species' former geographic range causes the animals stress.[11] This view is philosophically reasonable, internally consistent, and deeply compassionate.

Nevertheless, Artemisians adhere to deep norms as well, the relevant one being that society must not cause the regional or global extinction of any species for reasons of aesthetics, convenience, or profit. In the condor case, for example, I and my colleagues believed that by not intervening aggressively, we would be abandoning a fellow species, whose imminent demise was due primarily to human activities.

In the case of the wolf repatriation to the Yellowstone region, Artemisians and many Palesians believe that aggressive means are sometimes necessary to restore a species to its original homeland, particularly if the absence of the species is contributing to the degradation of Nature.[12] The wolf is a major ecological player—a keystone carnivore—and its extirpation from the Northern Range of Yellowstone National Park apparently contributed to the disappearance of an entire habitat (the beaver pond ecosystem in the Northern Range) due to elk overbrowsing of riparian vegetation, and to the suspension of ecologically critical natural processes such as the reproduction of aspen.[13]

Nevertheless, the Artemisian justification for intervention can be interpreted by Daphnian philosophers as granting greater value and privilege to the ecosystem or to the group (species) and to its survival as an evolutionary lineage than to the psychological well-being of some of the individuals that constitute the group at any given moment.[14] On the other hand, the Artemisians will argue that to nurture the survival, existence, and the possibility of future evolutionary change of a species is just as ethical as is respect for the dignity and freedom of currently living animals. There is no litmus test that allows us to decide objectively which view should prevail in a given situation. It might be worth considering that difficult decisions, whether made from within a Palesian, Daphnian, or Artemisian system of values, are based on deep-seated systems of belief and compassion that we can all respect, even as we disagree about their application.

DIMENSIONS OF WILDERNESS MANAGEMENT

The issues that we see most frequently in discourses on the management of wilderness include (1) the need to reverse processes of ecological degradation that were entrained by land use abuses and errors in the past, (2) the need for clear, measurable goals, (3) the duration of the management intervention and its visual (sensory) consequences for human beings, (4) the spatial extent of the intervention, and (5) the need for humility in light of past management mistakes. After briefly discussing these issues, I will illustrate their complexity using a current debate on the appropriateness of mechanized habitat restoration in a designated wilderness area in New Mexico.

The Human Imprint from the Past

The ecological consequences of historical impacts such as past land use practices is a frequent cause of debate. Historic insults or wounds vary from planetary catastrophes to minor, local perturbations. An example of a catastrophic impact is the extirpation of the megafauna in North America mentioned above. In much of Canada north of the 50th parallel (above Newfoundland, Winnipeg, and Vancouver), this perturbation may be the only serious human impact, though recent commercial logging and exploration for oil and gas is locally significant. Farther south, designated wilderness areas are raw with many additional wounds, including the effects of severe overgrazing, the recent extirpation of carnivores (wolf, grizzly bear, wolverine, Canada lynx, fisher, jaguar, cougar) and ungulates (bison, elk, bighorn sheep, pronghorn antelope) extraction of fiber, meat, and minerals, and a century or more of fire suppression, road-building, dam construction, and, more recently, the marks of technological recreation including mountain-bike and all-terrain-vehicle tracks, pitons in the rocks, and the less noticeable but measurable impacts on wildlife of river boating, snowmobiling, and backcountry skiing. As discussed below, some wilderness areas, if left alone, will continue to degrade because of entrenched positive feedback processes.

The Need for Clear Goals

When considering active management in wilderness areas, the onus must be squarely on the managers, scientists, or activists to clearly articulate their ecological goals. The targeted conditions must be reasonable and attainable. The end-point of restoration actions must also be quantitatively defined, and methods for evaluation, monitoring, and adaptive management, if any, must be spelled out.

The Time-Scale of Management Interventions

An important variable is the interval of time over which restorative manipulations occur. How long should we tolerate an invasive intervention for the sake of long-term benefits? For example, are the visual or ethical effects of seeing a bear with a radio collar or tattoos in its ears, or seeing ground-level stumps in a thinned ponderosa pine woodland, so egregious that they should not be tolerated for ten or twenty years, even if these interventions are essential to restore a high degree of naturalness to a wilderness? The burden of proof should always be on those who propose to manipulate designated wilderness.[15] Daphnians, Artemisians, and Palesians generally agree on this. The main point is that the interval during which invasive interventions occur should be minimal.

Size and Management

There are two relevant aspects of size. One issue is the relative or absolute size or extent of a proposed management activity, particularly ecological restoration.[16] The larger the wilderness, the larger a given intervention can be without doing violence to its autonomy, solitude, and to people's experience of the wilderness area as a whole.

The second issue is the size of the wilderness per se; big areas, particularly if they are physically linked, are less likely to require invasive management than small ones. This is because "edge effects" and "island effects" decrease in proportion to the size of a protected area.[17] Edge effects are ecosystem changes that occur around the periphery of wildlands. In general, the smaller a wildland, the greater the amount of edge habitat relative to the better buffered "interior." Edge effects include forces or phenomena that penetrate from the external world, including poaching, changes in the microclimates near the edge of a forest, invasions of weeds and pathogens, human-caused fires, pesticide and herbicide drift from farms or plantations, and harm to ground nesting birds and bird nests by edge specialists such as cowbirds, crows, magpies, raccoons, feral or domestic pets, and by livestock. In the tropics edge effects may penetrate up to ten kilometers into a preserve.[18]

Island effects are those dissipative processes that are exacerbated by smallness and isolation. The best known island effect is the gradual disappearance of species in newly isolated habitat remnants; the rate at which species disappear in isolated remnants is inversely proportional to size. The most vulnerable species are large animals and habitat specialists because their numbers are too low to survive random environmental and chance demographic events, not to mention the effects of inbreeding. Habitats disappear too, in part because large-scale processes—such as hurricanes, fires, and floods—that maintain the natural mosaic of ecosystems can affect the entire remnant, and will tend, therefore, to homogenize a small wildland. In a large wilderness area, such as Idaho's River of No Return, even major, stand-replacing fires are considered to be beneficial. In a small, isolated wilderness, however, like the Mesa Verde in Colorado, such fires can be overwhelming. Similarly, a single large storm can also devastate a relatively small wilderness.

Bigger land area is better for evolutionary processes as well, which, contrary to popular belief, are now at a standstill for large animals restricted to national parks and wilderness areas. The problem is that small wildlands have small populations, and natural selection is relatively powerless against genetic randomness when there are less than a few thousand individuals. Moreover, the smallest area in which a species of small vertebrate, such as a rodent or rabbit, can evolve into two new species by geographic isolation is about 110,000 square kilometers—the size of Cuba or Tennessee.[19] Larger animals require an area about the size of Madagascar (about 600,000 square kilometers) to speciate, which equals the combined areas of New Mexico and Arizona.[20] For comparison, only about 12 percent of designated wilderness areas in the United States are larger than one thousand square kilometers.[21]

Ideally, wilderness areas should be large enough for evolution to occur.[22] Sadly, though, the small size of most wilderness areas in North America south of the 50th parallel precludes this possibility, at least for critters equal to or greater than the size of a badger. Thus, to assume that the current set of designated wilderness areas in the United States can be crucibles of evolutionary self-renewal for nature is a delusion, though in the short run such areas may have the appearance of being self-willed or untrammeled. The obvious solution to these ecological and evolutionary size constraints is to change the management protocols on most public lands in North America and to create a much larger, better connected network of wilderness areas.[23]

The Need for Humility

One of the strongest arguments against management in wilderness is the history of disasters caused by past management activities, though most of these horror stories are more relevant to range lands and forest reserves than to wilderness, per se. Those opposed to management interventions have a compelling prima facie case. There is space for only one example. Early in the twentieth century, U.S. government scientists were seeking a biological control for gypsy moths, a serious forest defoliator. They eventually discovered a European fly (*Compsilura concinnata*) that happily parasitized the gypsy moths by placing eggs in the body of the caterpillar.

Sadly, it was recently discovered that the severe declines in hundreds of other insects in North America, including the mysterious giant silk moths, are likely to have been caused by the fly.[24]

Even though agency scientists are now trained to think more ecologically, the potential for serious mistakes cannot be discounted. Risk aversion should be the prime directive, even while accepting the argument that the worst policy sometimes, is doing nothing.[25] "These are questions of trust not only about science, however, but also [about] the people who apply it: scientists and land managers. When people oppose manipulative restoration, is it the science they distrust or is it the managers and the agencies they represent?"[26]

The preceding five issues all point to the use of the precautionary principle. In this context, the principle could be stated this way: don't mess with wilderness areas if there is more than the slightest chance of unforeseen, dire consequences. Another caveat refers to self-interest: don't support activities in wilderness if there is any evidence that the project is being proposed to enhance anyone's career, reputation, or status. This caveat also counsels against politically motivated management directives from higher-ups such as salvage logging or fire suppression that benefit business interests. Too often, wilderness management projects are promoted to maintain or enhance the size of an agency's budget.

THE BANDELIER CASE

How can this framework of value systems (Table 1) and the five criteria mentioned above be applied in designated wilderness? Consider the Bandelier National Monument in New Mexico. More than 90 percent (thirty thousand acres) of the Monument is managed as backcountry wilderness. Virtually all of the piñon-juniper forest in the monument— 11,730 acres—is within designated wilderness. The managers of the monument are responsible for the conservation of 2,500 known cultural (archeological) sites that are in this forest. Most of these sites are threatened by extraordinary rates of soil erosion caused by the anthropogenic loss of ground cover, in violation of the dictates of the National Park Service Organic Act.

Conditions in much of the piñon-juniper forest are rapidly deteriorating. In the past, the soil was protected by a well-developed herbaceous-grassland understory in the fire-sensitive piñon-juniper and ponderosa pine savannas and by open grasslands. These herbaceous habitats were self-maintaining because they allowed low intensity fires to spread frequently and to kill seedling conifers that otherwise would out-compete the understory plants and ground cover.

Several trends and events have caused unprecedented recruitment of piñon and juniper trees, leading to the closure of the canopy of the forest and the inhibition of forbs and grasses. The first trend may have been the disappearance of the Ancestral Puebloan (Anasazi) society from the region whose tree-cutting favored herbaceous vegetation. The second trend was the introduction of domestic livestock beginning around 1600. Grazing led to the virtual disappearance of the fire-conducting ground cover, and the suppression of widespread surface fires permitted the survival of many more young trees. Later, the federal policy of fire suppression initiated in the early 1900s had a similar effect. The third factor was the occurrence of severe droughts in the 1950s. Together, these factors contributed to a decrease in water infiltration, an increase in surface runoff from monsoon summer rains, and a decrease in fire frequency. Fires can no longer be carried through the habitat, thus allowing higher rates of tree seedling survival. Other factors militating against the re-establishment of a fire-conducting herbaceous ground cover are an unnaturally large elk population, due, in part, to the absence of wolves, and the abundance of seed-eating harvester ants.

Because the piñon-juniper trees are effective competitors for water and nutrients, an irreversible, positive feedback process is entrained, favoring less ground cover, closure of tree canopies, and more tree invasion. Bare ground under the trees has replaced herbaceous cover in a majority of sites, leading to high rates of runoff and the loss of topsoil at a rate of about one-half inch per decade. "To a significant degree, the park's biological productivity and cultural resources are literally washing away."[27] Unless humans intervene, the ecosystem will literally slide downhill, eventually reaching a relatively stable state characterized by low productivity, loss of vegetative diversity, and the disappearance of cultural heritage.

The obvious remedy is the use of chain saws to thin trees at ground level and to leave scattered slash to create a microclimate conducive to germination and survival of grass and forb seeds. Test plots have already shown that such a one-time treatment can restore the grassland in three years. Unfortunately, non-motorized means of thinning and limbing would require camps to be established for large numbers of workers, thus causing even greater impacts in the wilderness.

The restoration proponents have a clear and measurable goal: They conclude that "the end point should be achieved when there is sufficient herbaceous cover to carry naturally occurring fires."[28] This project addresses a demonstrably deteriorating situation. The question is, Is a short-term but mechanized restoration project in designated wilderness justifiable to protect "naturalness," ecological integrity, and cultural values?

It will require about two decades for signs of restoration (stumps and slash) to disappear. Should supporters of wilderness tolerate such mechanized means of restoration, assuming this would be the only practical way to reverse the degradative processes? Most ecologists and managers would say "yes" because their vision favors long-term ecological values and the healing of anthropogenic wounds, and because the project addresses the five issues of concern described above. Some Daphnians, however, might see the trade-off as too costly in terms of the autonomy and solitude criteria of wilderness.

CONCLUSION

The Bandelier example highlights the continuing relevance of the human "imprint" or "trammel" criterion of wilderness.[29] But imprint or trammeling is in the eye of the sojourner. To someone trained to recognize the signs of past or current overgrazing, for example, wilderness areas like the Bandelier wildlands are too "sick" to recover on their own. On the other hand, to the wilderness aesthete who is deeply upset by signs of human hegemony and control, the sight of fresh stumps in thinned forests is just as offensive as an ecological wound to the ecologist. As Jack Turner notes, "Or look at our ability to track snow leopards by satellite...Look at radio collars for everything from minnows and frogs to trout and moose...We must learn to see this 'control' not literally, as Orwell's television cameras with Big Brother watching, but more along the lines of Foucault's analysis of power—forms of power and manipulation so diffuse that we barely notice them."[30]

It is this difference in values and taste that is at the root of the argument about wilderness management. Ecologists tend to judge wilderness by the quality and fullness of the biota, now and into the distant future, and they tolerate less than ideal means to achieve quality and fullness. I don't like seeing collars on wolves and grizzlies either, but I think it a small price to pay—for us and the animals—if this is the only way that these grand creatures can be repatriated to a truncated land community in the current political climate. In fact, to stand in the way of wilderness restoration of this kind, using the autonomy argument for support, is to replace one kind of tyranny—pragmatic and scientific—with another form of philosophical elitism. Similarly, a few months of chain-saw work in Bandelier's designated wilderness, though an insult to solitude and to a time-bound view of trammeling, will likely reverse the slide to oblivion. Is it justified? I think it is, but our Daphnean philosophers would remind us of C. P. Snow's warning about stepping onto a moral escalator: it may be hard to get off.

One of my mentors is Arne Naess, a Norwegian mountaineer, amateur boxer, celebrated philosopher, and creator of the deep ecology movement. Arne lives for the wild, but his idea of the wild, akin to E. O. Wilson's and Gary Snyder's, includes little things. I once made the mistake of taking Arne on a backpack trip to the Grand Canyon. The grandiosity of the canyon was overwhelming to Arne's sensibility. So we left the canyon after only a day and ended up in a little, dry desert arroyo surrounded by small, brush-covered, gray hills. Arne spent his time there writing, walking around, and inspecting on hands and knees the tiniest flowers in the desiccated riverbed. I'm not saying that Arne's way of loving the wild is the best. I happen to like grand vistas. But I love lizards and flowers too. The point is that there are many paths of inner wildness and outer wildness. And there are many paths to inner and outer freedom. The nature-haters are gleeful when we fight amongst ourselves about which path is most pure.

❧

I thank Dave Foreman for allowing me to glean some ideas from his forthcoming book, THE WAR ON NATURE, *and Ted Kerasote and Dave for their helpful suggestions and comments.*

Marketing the Image of the Wild
Hal Herring

⟶ It's a cold November morning in the southern Bitterroot Mountains of Montana. Overhead the stars glitter, snow lies drifted in the gullies and sculpted along the ridges, and to the east the dawn sky lightens behind the jagged line of the mountains. In scattered bands, elk are moving down from the high country to their winter range.

Each herd is led by a seasoned cow, wise in the ways of coyote and wolf, lynx and lion, and human. They keep to trails worn deep into the ridges by generations of their passing, and there is constant chatter—mysterious whistles and sharp harsh barks—floating up and down the line of travelers. Flanking the cows, young bulls move through the timber like ghosts, wild-eyed and gaunt from the last of the breeding season. Plastered with rank-smelling mud from wallows, they're now veterans of battles fought in high-country meadows and witnessed only by ravens and wolverines.

The oldest and strongest bulls have remained behind, in the roughest country, healing and recovering alone, their genes secure in the bands of pregnant cows. When the snow gets so deep that they can barely travel, they might follow the old trails down...or they might not. Some of them will simply die in the high country—of old age, accumulated wounds, or exhaustion.

The band arrives at a place where the timber gives way to scattered juniper and fir, the ridgeline falling away on a steep south face covered with bunchgrass, sage, and wild rye. The elk stand, ears cupped, noses to the wind, their hides—sun-bleached tans and rich mahogany—a camouflage of sunlight and shade. The high country, the rich green thickets born of last year's snows, the pounding thunderstorms and incandescent alpine light, are now far away. The elk have accomplished yet another passage between worlds.

Animals like these—graceful, wary, and nomadic—have represented the freedom of wild country since the first Europeans settled in North America. The literature of wilderness hunting is replete with tales of pursuing wapiti in remote country like the Yellowstone, the Sangre De Cristo Range, the Gros Ventre, or any of a thousand places across western North America—places whose names evoke visions of some of the most pristine country left in the world.

Over the course of the last decade, a rapidly expanding "elk ranching" or "game farming" industry (it goes by both names) has exploited that very symbol of wildness, playing off our collective vision of the purity of wild country, to sell antler velvet to consumers around the world, and trophy shooting experiences in North America. The expansion has been accompanied by controversy every step of the way, no more so than in what many consider "the last best place."

In the election of 2000, Montana voters passed an initiative called I-143, which not only banned the most lucrative aspect of the elk ranching industry—the shooting of domestic elk bulls and other captive big game animals in fenced enclosures—it also prohibited the issuing of any new game ranch licenses. The same voters chose a Republican governor, Judy Martz, who promised a return to the old ways of extractive industry, and re-elected for a third term Republican Senator Conrad Burns, the avowed advocate of ranchers, loggers, and industrial entrepreneurs. The history of this dichotomy—how such a restrictive and liberal-minded ballot initiative took root in such conservative soil—is indistinguishable from a history of the game farming industry itself.

It is a tale about warnings that went unheeded, threats realized, and the vision of a future where wildlife is relegated to a commodity, confined to feedlots, and trotted out periodically to be killed in a grotesque, virtual-reality parody of the ancient wild relationship between hunter and prey.

A BURGEONING INDUSTRY

At the time I-143 passed by a majority of 52 percent, there were eighty-seven game farms in Montana, and about sixteen of them offered their clients the opportunity to shoot a domestic animal, usually a big bull elk, inside their fences. Many game farms, even those used for trophy shooting, are quite small—one enclosure near Whitefish, Montana, is only eighty acres, while others in the state involve the fencing of several thousand acres.

The number of game farms in Montana mirrors the growth of the industry across North America, despite current setbacks caused by a proliferation of wildlife diseases. According to the North American Elk Breeder's Association (NAEBA), an industry trade group, there are about three thousand elk farms in the United States and Canada, holding between 150,000–160,000 domestic elk behind eight-foot high "game-proof" fences. Only Wyoming and California have prohibited the industry entirely, and in Wyoming the legal challenges to that prohibition threatened to empty the state's coffers. The economics of a game farm are fairly simple. Although there is clearly a market for elk meat, animals have so far been too valuable for the slaughterhouse. Beyond the brisk trade in breeding stock, most profits come from mature bull elk. A bull elk produces a new set of antlers every spring, and during this growth stage the hard, ivory-like material typical of mature antlers is absent. Instead, they're composed of soft, extraordinarily blood-rich tissue—velvet—which has been long prized in the Orient as a restorative tonic and aphrodisiac.

In recent years, producers have also created a demand for velvet in the supplemental and alternative-medicine market of North America. Velvet currently sells for between $20 and $70 per pound, and is harvested annually during the summer by sawing off the complete rack of the bull. Ever since gory photos depicting this process were published in a 1989 issue of *The Albuquerque Tribune*, "velveting" has been carefully guarded from the public's eyes. By age seven, a good bull can produce as much as forty pounds of velvet. In addition, every September after the bull reaches fifteen months of age, a rancher can use an electrical ejaculator to harvest as many "semen straws" as the animal can produce. They can bring up to $750 apiece from a superior bull, but overharvest of semen reduces the amount of velvet produced the following year. When production of the bull's velvet slows due to age, he can be turned out into a shooting enclosure, where a client will pay anywhere from $5,900 to over $20,000 to shoot him and take home his antlers and perhaps his meat. Using the semen straws, a rancher can continue to breed his prize bulls long after their heads are mounted in a client's living room.

A QUESTION OF VALUES

Reducing elk to this sort of engineered commodity inspires visceral revulsion in people who revere wildness in wildlife. Not surprisingly, the most vehement, and the most effective, opposition to the elk ranching industry has come from the ranks of traditional big game hunters, for whom elk are the flesh-and-blood symbols of unsullied ecosystems and all that is right with American hunting: the ideal of strenuous pursuit coupled with advocacy for wildlife and wildlands preservation.

One of the most outspoken critics has been Gary Holmquist, a Montana hunter and retired Marine Corps colonel, who led the fight to ban captive trophy shooting in Montana. "I have plenty of environmental reasons to oppose this industry," he said, "but my primary reason has always been an emotional one: I see these magnificent animals, with their horns cut off, standing around in the mud of the feedlot, staring through the fence. And then the rancher turns around and sells them as a trophy of the hunt? It is disgusting, and it debases both the animals and the concept of hunting."

However, Steve Wolcott, a leading member of the North American Elk Breeders Association who raises elk in Colorado, finds Holmquist's critique shallow. "They hate us because we domesticate an image that they have," he says flatly.

Another elk rancher and former board member of NAEBA, Bob Spoklie, adds, "We should have the right to harvest our livestock and sell it to the highest bidder, just like any other producer. As far as I'm concerned, we lost fair chase when we gave the Indians a horse and a rifle."

For Marty Boehm, a third Montana elk rancher, the controversy over elk farming isn't only about one group of people trying to force their perceptions onto another. It's about economics. "I have fifty acres, and I can tell you that there is nothing else legal that I can do on that ground that would allow me to survive."

The captive trophy shooting business is also booming in states like Utah, Colorado, and Idaho, which have no laws restricting the practice. "We do license elk ranches," said Dr. Phil Mamer, a veterinarian for the Idaho Department of Agriculture, "but we don't have any rules about how you harvest your animals. You can shoot your livestock if you want to."

There is little doubt that the market for domestic trophies is driven by the ever increasing hunting pressure on both private and public lands, where big game animals, at least in those areas accessible by roads, do not survive long enough to attain maturity. International big game hunter Bill Dougherty, who works for a Texas-based company that sells state-of-the-art fencing for game farms, firmly believes that fenced shooting operations are the future of hunting. "I have mixed emotions about it," he adds, "but probably 50 percent of the hunting I do is behind a fence. I don't hunt public land anymore. It's dangerous, and the animals get killed as soon as they have legal horns."

Dougherty goes on to explain that there are many potential clients of shooting operations who value the idea of the hunt, but who are unable, or unwilling, to pursue game animals in traditional ways. "Most people who have the money to hunt don't want to get on a packstring and travel through the wilderness in the snow," he says. "The numbers of people who want to do that are dwindling, while the numbers of people who want to take a quality animal are going up."

Dougherty recently shot a trophy bull elk on a fenced ranch in Utah, using a black powder rifle. "It took me two and a half days to get close enough to the animal that I wanted," he said. "I call that hunting, and I enjoyed it." He says that the client's experience depends on how the ranch is managed. "A lot of ranches are playing the whole game—artificial insemination, velveting, selling breeding stock, and then selling hunts. The bulls get handled a lot, and the hunts become very predictable. Where I hunt, the animals are just turned into the enclosure, and then they're gone."

High out-of-state license fees—$450 in Montana and $410 in Wyoming—have also helped the game-ranching industry, since no license is required to shoot on a game farm and the results are guaranteed: an elk whose rack is as big or bigger than anything found in the wild. However, by not purchasing a hunting license, game-farm hunters contribute nothing to state funds earmarked for wildlife conservation and habitat purchases. Another reason game ranching has thrived is that fenced habitat, at least in the Rocky Mountains, gives the impression of wild country. In fact, not that long ago much of it was wild country. A measure of the industry's sophistication is that many shooting enclosures are now constructed in the rough, natural habitat that elk and deer favor. Elk ranch consultant Craig Hayes, of Paso Robles, California, helped design the shooting enclosures on two successful Montana operations and says, "Our most scenic ranch is the Royal Pine, near Bigfork. We do mostly horseback hunts there, and you look out over the Swan Range on one side, and Flathead Lake on the other." At the other Montana ranch where Hayes worked—the Big Velvet Elk Ranch, near Darby—most shoots end with a video and photos of the client astride a trophy bull in the sagebrush, with the snowy, ten-thousand-foot summit of Trapper Peak in the background.

There is a high price to pay for such postcard scenery and snippets of wildness. In almost every state, native big game cannot be fenced in by landowners since, by law, private citizens in North America are unable to own wildlife while it is alive. On the hoof, or on the wing, it remains a public resource, managed by the state or province, until it's killed by an individual holding a valid hunting license. Only then can the public resource be converted to private property. Consequently, whenever a game ranch erects a game-proof fence, the public's game animals must be displaced either by hazing or eradication. The landowner then restocks his or her land with, most commonly, domestic elk or deer.

The state of Texas is an exception to this rule. There, landowners can erect a game-proof fence without removing native white-tailed deer. The animals become the de facto property of landowners, who then charge a trespass fee, usually in the form of a lease, to allow hunter's access to what remain, on paper, public property. Texas is also the leader in providing fenced trophy shoots for "exotic animals" such as axis deer and greater kudu. (In the terminology of wildlife biology, "exotic" means not indigenous.)

When accused of evicting the public wildlife in this way, game ranchers in states other than Texas retort that the amount of land appropriated and wildlife removed is too minuscule to matter. However, that isn't necessarily the case. When the fences went up for the Big Velvet, in 1993, Len Wallace, the owner of the ranch, hired local teenagers to run through the area, hazing the scattered herds of mule deer and whitetails towards the fence perimeter. Helicopters assisted by buzzing the animals. The area was then opened to public hunting, and finally game wardens stepped in to shoot forty-nine mule deer that had evaded the removal effort. Wallace later applied for a permit to fence another 1,100 acres of his property, which would have involved displacing an estimated 750 mule deer and a small herd of resident elk. "There is absolutely no benefit for me in feeding the public's wildlife," he explained, "and I have no intention of doing so." Montana wildlife officials denied the permit, citing the negative impact on public hunting opportunities that would result from the loss of so much winter range as well as "overwhelming negative public opinion."

At least in Montana—despite strenuous lobbying efforts by the elk ranching industry to place themselves solely under the far more lenient regulatory power of the Department of Livestock—wildlife officials still share responsibility for issuing permits and monitoring game ranches. In states like Colorado and Idaho, the industry is regulated only by the states' agricultural agencies. "We review game ranch proposals," says Mike Miller, of the Colorado Division of Wildlife, "but I can't think of a permit that has ever been denied because of wildlife concerns. The Department of Agriculture is directed toward promoting the industry, and they are very reluctant to turn down applications."

Across the United States and Canada, the exact acreage now enclosed by eight-foot-high game-proof fences is unknown. According to one critic of the industry, Darryl Rowledge, of the Alliance for Public Wildlife, in Calgary, Alberta, "If the elk ranching industry actually achieved their long-term goals for expansion, it would constitute the largest outright loss of wildlife habitat ever seen in North America."

HYBRIDIZATION AND DISEASE

Ironically, the "game-proof" fences required to keep elk and deer in captivity have proved all too permeable. In Montana alone, dozens of elk and deer have escaped, while native elk, deer, bighorn sheep, and black bears have gone the other way, seeking food, mates, prey, and companionship. At last count, late in 1999, at least seventy-five domestic elk have escaped into Montana wildlands and have not been recovered.

Such escapees worry state wildlife officials, because extensive testing of elk on Montana game farms over the past two years has turned up forty animals that are crosses between Rocky Mountain elk and European red deer. The latter have been domesticated in Europe and New Zealand for many generations and look very much like elk but are easier to handle and exhibit more vigor in captivity.

Whether they would have any significant effect on wild Rocky Mountain elk if they were to breed among them remains unknown. According to Tim Feldner, the game ranch specialist for Montana Department of Fish, Wildlife, and Parks, "We have what we believe is a genetically pure pool of Rocky Mountain elk, adapted over thousands of years to this land and this climate. Nobody can say what effects we could see if red deer genes were introduced."

But it isn't genetic contamination from escapees that most worries wildlife advocates these days. From the first days of elk ranching, biologists have warned that gathering together large groups of elk from many different places of origin and keeping them in confined spaces would create breeding grounds for a variety of diseases, which could then be passed on to wild elk. It didn't help that so many game farms were being established directly in the habitat of wild elk herds, and that game farmers had rejected requests from wildlife officials to install expensive double fences to prevent wild elk from making nose-to-nose contact with their captive counterparts, one of the surest ways of transmitting diseases. The biologists' concerns haven't been misplaced.

Cinnabar Game Farm, situated just north of Yellowstone National Park, in Montana's Paradise Valley, started business in 1946, marketing both live elk and velvet to Asian buyers. Its owner, Welsh Brogan, expanded his enclosures in the 1970s and came under fire from local outfitters and hunters who claimed that his fences blocked antelope migration on one side of the Yellowstone River. That claim proved to be only the opening volley of charges made against the Cinnabar. As Jim Kropp, a captain of game wardens for the region, told a reporter for *The Missoulian*, "The ranch grew so big, so fast, and it was right up against the Absaroka-Beartooth Wilderness, which is habitat for thousands of wild elk. Wild and domestic elk could literally rub noses through the fence. All of a sudden the potential for disaster became huge."[1]

Concern escalated when bovine tuberculosis—a disease that is transmissible to humans—appeared in a shipment of elk that Brogan sold to a Canadian game farm. The Cinnabar was placed under a five-year tuberculosis quarantine, and the following year Brogan was charged with the illegal capture of eighty wild elk. After a lengthy court battle, he was found guilty and in 1991 lost his license to operate the game farm.

The specter of bovine TB did not disappear, however. Over the next four years, the bacillus would spread through confined elk herds. Six more Montana farms were placed under TB quarantine, and at one of them, the Elk Valley Farm near the town of Hardin, a wild mule deer and a coyote killed outside the fences were found to have been infected. At Royal Elk Ranch, near Gunnison, Colorado, domestic elk tested positive for the disease, and the entire herd was slaughtered and incinerated on site. Wildlife officials could only wait and hope that the TB would not jump from captive to wild elk. As Dr. Mitchell Essey, the senior staff veterinarian for the USDA, said, "I don't know what we will do if TB gets established in wild populations. No one knows how we would control it if it got into elk herds like those in Yellowstone National Park. The potential ramifications are almost inconceivable."[2]

In Canada, the situation proved far worse than in the United States. The industry had been expanding very rapidly, the trade in breeding stock frenetic. Unknown to ranchers, it was a trade in TB infection as well. After discovering TB in game-farmed elk, Canadian livestock officials struck hard, "depopulating" fifty game farms by slaughtering 2,500 infected or exposed elk. Forty-one people were treated for exposure to the disease, and, in a decision that remains controversial a decade later, Alberta's elk ranchers received over $13 million in compensation for their destroyed herds, even though the Canadian government had permitted them to establish a portion of their breeding stock through the capture of wild elk, which of course were public property.

Despite these crises, elk producers are extremely proud of the proactive and apparently successful way that the bovine TB crisis was handled. Better regulations and herd monitoring have indeed reduced the risk of the disease. Also, it was discovered that the original test for bovine TB, developed for cattle, was almost completely ineffective when used on elk. Infected elk were being traded among ranches across the continent after testing negative for the disease. A new test has proved much more reliable.

Unfortunately, as the echoes of the TB problem have faded, a new threat has emerged. Chronic Wasting disease (CWD) was once known in only a small contiguous area of northeastern Colorado and southeastern Wyoming, where it infects wild mule deer and elk. In 1996, it appeared among domestic elk in Saskatchewan. Once again the elk industry was in a period of rapid expansion, and the infection traveled in shipments of breeding stock. Game farms in Colorado, South Dakota, Montana, Nebraska, and Oklahoma were eventually infected.

CWD is a member of the family of transmissible spongiform encephalopathies (TSE) that include mad cow disease, scrapie, and Creutzfeldt-Jacob disease (CJD). For centuries, scrapie has periodically swept through domestic sheep herds, passed among them through contaminated birth fluids, usually while the animals graze on pastures where lambing has occurred. No evidence exists that scrapie affects humans. The human form of TSE, Creutzfeldt-Jacob disease, strikes at the rate of about one per million individuals and takes years to present symptoms. Researchers remain unsure as to the nature of TSE's causative agent. Some think it's a virus, others a virino (a small nucleic acid surrounded by a protein), still a third group believes it to be a prion (a modified host protein).[3]

In all cases, TSE infections result in a degeneration of brain tissue, giving rise to the "spongiform" appearance that gives the conditions their name. Symptoms vary from the slobbering and staggering of deer and elk that have chronic wasting disease to the forgetfulness and progressive dementia of humans with CJD. All TSE infections, as far as is known, have proven fatal.

It is generally accepted that the mad cow epidemic in Europe was a human-caused phenomena, resulting from the rendering of sheep and cattle parts, some of them infected with the TSE agent, into high protein cattle feeds. As the mad cow crisis developed, a new form of Creutzfeldt-Jacob disease began to appear, affecting much younger people than the traditional form and with a much faster rate of onset. Called new variant Creutzfeldt-Jacob disease (nvCJD), it has so far killed ninety-six people in Europe, and has been linked to the consumption of TSE-infected beef products. At present, there is little evidence to suggest that consumption of game meat contaminated with Chronic Wasting disease has ever created nvCJD in human beings, but the possibility has not been ruled out.[4]

No one knows how Chronic Wasting disease is transmitted, or exactly what the incubation period for the disease might be. CWD was first found in captive mule deer at a research facility near Fort Collins, Colorado, in 1967. The deer were part of a forage study being conducted by the Colorado Division of Wildlife, and as Beth Williams, a professor of veterinary science at the University of Wyoming, and one of the foremost experts on CWD, says, "We don't know where the disease came from. Research animals originally came from the wild, and it's impossible to tell at this point whether CWD arose in those facilities or in native herds."[5]

What is certain is that CWD has reached levels of infection in domestic elk that have never been seen in the wild. Another expert on the disease, Dr. Tom Cline, a veterinarian with the South Dakota Animal Industry Board, observes that "captives stay around much longer, and they have a much greater chance to pass the disease around. If you've got a mule deer or an elk staggering around in the wild, a predator will just take it down, probably before it has a chance to pass the disease on."

One of the mysteries of TSE infections, and one of the reasons that they have been so difficult to investigate in domestic animals, is that although the infectious agent causes severe breakdown of the brain tissues, it triggers no response from the victim's immune system, which makes finding a test for the disease in living animals very difficult. Despite over $200,000 spent in research efforts by the elk industry, so far there is no way to test live animals for infection.

Canada is now in the throes of a severe outbreak of the disease among domestic elk, which officials traced to shipments of animals from the United States. As of April of 2001, nineteen game farms in Saskatchewan have been depopulated, 3,026 elk have been slaughtered in an attempt to arrest the spread of the disease, and compensation payments—again coming from the public's coffers—have ranged up to $4,000 per animal.

"The taxpayers are paying through the nose," says Dr. Valerius Geist, a professor emeritus of environmental science at the University of Calgary, "for an industry which is serving as a disease bridge to the public's wildlife. And if that bridge is crossed, we will have to present some quick answers to the problem, or they will begin slaughtering our wildlife." Geist, who has been warning the Canadian government for years that CWD would one day show up on domestic elk ranches, is currently organizing sportsmen's groups to address the threat of CWD spreading from game farms to the wild.

Wildlife officials in the United States and Canada remain on watch and hope that CWD will not become established among deer and elk in states where it currently does not exist. If it does, there appears to be nothing else to do except slaughter nearby wildlife. Mike Miller, of the Colorado Division of Wildlife, has said, "Trying to manage CWD in wild herds is hell if you do, hell if you don't. About the only way we know how to manage it is to do a density reduction on the animals that may be carrying it, which basically means clobbering the native wildlife."

In Colorado, this stark way of controlling CWD in wild herds has already been initiated. During the winter of 2001, wardens attempted to reduce the density of deer and elk in the northeastern corner of the state where CWD is endemic. A special hunt was held, with hunters informed of the possible risks, and asked to submit the heads of their kills for testing. Out of 206 deer killed, nineteen turned up positive for CWD, according to Todd Malmsbury, a spokesman for the Colorado Division of Wildlife. "That figure is not quite representative of the worst area," Malmsbury added, "because most of our hunters stayed to the south, where there is less of a problem. We have wardens that are going to be shooting another hundred or so animals along the Wyoming border, and we will continue our efforts to reduce the herd over the whole area by about 50 percent, which should slow the spread of the disease."

Even with these recurring outbreaks, the elk industry's website, *www.naelk.org*, has no information on the CWD epidemic in Canada, nor does it mention what has become of the CWD-infected ranches across the western U.S. Instead, in a position statement concerning the disease, it claims that free-ranging wildlife is the major vector for CWD and poses a threat to the future of the industry: "After farmers and ranchers spend millions of dollars and years of sacrifice to eliminate a disease, they are faced with the threat of re-infection from wildlife whose managers refuse to act responsibly to control the same disease." One can only infer that such responsible action would mean the eradication of the public's wildlife.

Commenting on that dire possibility, Darrel Rowledge, of the Alliance for Public Wildlife, says, "There is no rational justification for this industry. The bigger it gets, the more it costs. And for what? What do they produce? Elk velvet, 'the natural Viagra,' and sport shoots for domestic animals. How do you defend that in light of the threats posed to wildlife? The government focuses on solving these disease issues, without ever addressing the real problem, which is the industry itself."

THE RIFT WIDENS

It seems doubtful that voters in other states in the U.S. will follow Montana's lead in curtailing the elk industry, and perhaps with good reason. It isn't certain whether the Montana initiative will stand up to a recent legal challenge by game rancher Kim Kafka, who has sued the state's departments of livestock and wildlife, claiming $50 million in damages for the taking of his private property without compensation.

Market forces may also be kind to game ranchers. In 2000 the elk industry suffered a setback when Korea embargoed 24,000 pounds of antler velvet because of concerns over Chronic Wasting disease. The move sent velvet prices into a dive. However, with big-racked bull elk standing by unshorn, the captive shooting market became more attractive. Minnesota, for example, which has the most elk ranches of any state in the nation, is rethinking its prohibition on captive shooting, as is Alberta. The price for breeding stock has also fallen low enough to send at least a few elk to the slaughterhouse. In South Dakota, domestic elk carcasses must be held until the brain tissue has tested negative for CWD before the meat can be sold, but most other states do not require this caution. Special slaughter facilities for elk have been established in Denver, Colorado and in the Flathead Valley of Montana.

Against this economic backdrop, proponents of elk ranching claim that a domestic elk requires one third of the forage required to produce one domestic cow, and that elk enclosures are ideal ecological uses for lands that can't produce crops or provide profitable grazing for traditional livestock. There is another side to this argument: as human population rises dramatically in the Rockies, it is just these sorts of lands that are becoming more valuable as wildlife habitat, not fenced areas that exclude native fauna. However, as agricultural producers struggle with an economy that seems determined to leave them behind, it is likely (barring legal prohibitions) that the elk business will expand into these agriculturally marginal but wildlife-rich landscapes, diminishing the habitat of wild elk and confining them, along with other wildlife, to public lands that are increasingly eyed for energy development. The willingness of so-called sportsmen to embrace this train of events reveals to some hunters, like hunting ethicist Jim Posewitz, "an evil seed" buried in the heart of modern hunting—the idea that, somehow, hunting can be separated from the need to protect wild ecosystems and the attendant respect for the creatures with whom hunters share their world.

Other big game hunters, searching for a metaphor to explain their outrage at fenced trophy shooting, have compared the activity to prostitution. Both activities involve paying a price for instant gratification, and both serve as a simple, diminished substitute for an experience that is inherently complex on both a physical and spiritual level. The problem with this metaphor is that in the case of prostitution, the ugliness of its diseases and debasement affect only our species.

Imprecise though it may be, the comparison has served as a rallying point—the elk hunters of Montana worked for months to bring the game farming initiative before the voters, and the majority of them, urban and rural alike, agreed that marketing a denatured version of wild animals and wild places could not be tolerated, especially when the short-term economic well-being of a few jeopardized the health and safety of the many. In this regard, the game ranching controversy reflects a much wider clash of values that is currently dividing American culture.

One side in this battle believes that nature is only valuable in so far as it pays its way or can be manipulated to turn a greater profit. The other side declares that wildness not only provides clean air, potable water, and monetary returns, it also maintains our sanity and spiritual well-being. As human populations expand throughout the world, the clash between these two value systems will grow more intense. The outcome could not be more in doubt.

The Once and Future Grizzly

Todd Wilkinson

∽ At the turn of the twentieth century, William Henry Wright shot
dozens of grizzly bears in the northern Rocky Mountains. He stalked some
bears for weeks at a stretch—alone, trailing them dozens of miles through
forests and mountains before pulling within firing range. He shot grizzlies
in fishing holes where they fed vulnerably on plump spawning salmon; he
ambushed them when they came to prey on domestic sheep flocks and big
game; and he felled bears—old wary silvertips and curious yearlings
alike—for fun.

In Wright's most famous episode, which he proudly chronicled in his
classic but unnerving book, *The Grizzly Bear*, he slew five bruins in a single
setting, firing just five bullets from his Winchester. Likening the experience
to wingshooting ducks from behind a blind, he called the feat "the greatest
bag of grizzlies that I have ever made single handed."[1]

It's true that Mr. Wright made money as a hunting guide and from selling the hides of bears he killed. And he justified his actions by believing he
was doing citizens a public service, just as Buffalo Bill did with the thousands of bison he shot, all in the name of progress.

But the fact remains that what truly attracted Wright to his avocation
was the challenge of the chase, and the empowering feeling that came with
subduing something more powerful and majestic than himself.

Today, grizzlies are long gone from the rugged wildland complex
known as the Selway-Bitterroot in western Montana and east-central Idaho,
and William Henry Wright—a hero to some, a villain to others—might
best be assessed as a touchstone from which we can measure our powers of
redemption. Can we—humans in the richest nation on Earth—right an
environmental wrong that occurred in a less informed age and restore

ecologically degraded and "vacant" landscapes by returning their large aboriginal animals?

GROUND BREAKERS

In a conference room in Missoula, Montana, Hank Fischer, from Defenders of Wildlife, Tom France, an attorney with the National Wildlife Federation, and I look over a map showing the areas in nearby central Idaho where the first grizzlies might be released under a Fish and Wildlife Service plan.

Following on the heels of returning wolves to Yellowstone and central Idaho during the 1990s, the Service has taken an unprecedented step with grizzlies. For one, grizzly reintroduction has never been attempted before. Second, the Selway-Bitterroot plan stands out as a rejection of the top-down federalism that has defined most major conservation victories of the last century, including the controversial reintroductions of wolves across the United States. This is no small reversal, for it is the perceived "management from Washington"—most recently coming with a Clintonesque spin—that continues to fuel local backlash in the hinterlands of the West. The plan also shakes up people's assumptions about our ability to live with animals who can eat us, for it restores a large dangerous predator over a huge swath of country used by thousands of recreators.

The bear's new homeland would cover 21,645 square miles between the flanks of the Bitterroot Mountains on the east and the Clearwater-Salmon river drainages on the west. It's an area equal to the size of Vermont, Massachusetts and Connecticut, and, by design, it targets remote country—among the most geographically isolated in the contiguous United States. It includes all of the Selway-Bitterroot, Frank Church-River of No Return, Sawtooth, and Gospel Hump federally designated wilderness areas—places where industrial activity, motorized access, and human settlement aren't allowed.

"There isn't another spot like it west of the Mississippi," says Fischer.[2] "We're talking about an area of 7.8 million acres of roadless habitat, almost four times bigger than Yellowstone which has 180 miles of road and 3.3 million people passing through it. The Selway-Bitterroot is a place where grizzlies used to thrive, but now it is vacant of bears. There is a rightness about returning the bear here that cannot be denied."

France adds that the Selway-Bitterroot represents a linchpin that could help connect the "island" populations of grizzlies in the greater Yellowstone region to the south with bears in Glacier National Park, the adjacent Bob Marshall-Scapegoat Wilderness Areas (called by some "the Crown of the Continent"), and the Canadian Rockies to the north.

Between greater Yellowstone and the Crown just one thousand grizzlies are left in the lower forty-eight states. There might be eighty more grizzlies in the Cabinet-Yak mountains of northwestern Montana, the Selkirks, and the Northern Cascade region of Washington. This relatively small number of grizzlies is relegated to less than two percent of its original vast range south of the 49th parallel, and conservation biologists say that island populations of species—in essence populations of animals that are cut off from other populations of their own kind—suffer far greater rates of extirpation and eventual extinction.[3] "It is only through a major expansion of its distribution and abundance that the species can hope to achieve a level of long-term viability," writes conservation biologist Mark Shaffer. "Merely attempting to stabilize existing populations is too little, too late. The only way to keep the grizzly bear in the American West is to allow it to return to a larger portion of the regional landscape than it currently occupies and to reestablish links between populations."[4]

EVOLUTION OF THE PLAN

In keeping with its non-top-down style of management, the plan had an unusual beginning. At a 1993 Fish and Wildlife Service meeting, the agency announced that the Selway-Bitterroot was at the head of its list for areas in which to reintroduce grizzlies.

A timber industry lobbyist named Dan Johnson raised his hand and said, "We don't want the damn bear." In his next breath, Johnson made an astounding remark. Believing that grizzly reintroduction would occur whether he liked it or not—especially with Bill Clinton in the White House—Johnson challenged the bear bureaucrats in the room to consult with the timber industry if they were going to craft an effective plan for getting bears back on the ground.

Johnson's comment could have been dismissed. After all, the conservationists in the room had waged an ongoing war with timbermen over how much cutting should be allowed on national forests. Working with them was, in the mind of many enviros, tantamount to sleeping with the devil. But Hank Fischer and Tom France looked at each other and raised their eyebrows. Both realists, they knew that grizzly reintroduction in the Selway-Bitterroot was, at best, years away. In fact, given the conservative, anti-federalism politics of Idaho, it might never happen. The two men came to a decision. That night, as Johnson drank a beer in the hotel lounge, Fischer approached him and asked, "Mind if I join you?"

It proved to be a metaphoric question.

Accepting an invitation to meet two months later, Johnson showed up at France's office in Missoula with a van load of workers from the Potlatch Timber Mill. "In walked this group of husky loggers who all looked like they were offensive linemen for the Green Bay Packers," France recalls. "They sat there. No smiles, no love for environmentalists whom some of them blamed for the demise of the timber industry. But they came to listen."

"What would it take for you to support grizzly reintroduction in the Selway-Bitterroots?" Fisher and France asked the timbermen.

"What would it take for you to honestly work with the timber industry rather than against us?" came the reply.

Tired of the atmosphere of terminal bitterness that had divided environmentalists and loggers in the Pacific Northwest, Fischer and France saw an opening. They had learned painfully from the protracted, costly battle to bring back wolves. And they saw how the Endangered Species Act, thanks to an amendment passed by Congress in 1982, could be more flexible in addressing conflicts between ecology and economics. "We were looking for a cheaper, faster way of achieving species restoration," France said. "And they [the timber industry] didn't want to get waylaid by more government regulation."

The timber industry worried that the grizzly might emerge as another spotted owl and bring some logging operations to a standstill. Johnson, and a colleague, Jim Riley from the Intermountain Forest Industry Association, recognized that Fischer and France were sincerely interested in charting a new course. With nothing to lose, the two sides formed a pact based on trust that would be built one meeting at a time.

DEVILS IN THE NEW FOREST

Initially, several different conservation groups attended the nascent discussions, including the Alliance for the Wild Rockies, the Sierra Club, the Idaho Conservation League, and Friends of the Bitterroot, but most abandoned the talks because they felt the timber industry had ulterior motives, foremost among them weakening protection for endangered species. In addition, some of these activists favor eliminating all logging and livestock grazing from public land. "When you crawl into bed with the enemy, you become the enemy," warned Montana conservationist Steve Kelly.

Meanwhile in Idaho, a state fish and game commissioner named John Burns whipped up a frenzy in the town of Salmon using similar words to try to shame loggers who showed signs of going soft on bears and greens. "Don't just say 'No!' to grizzlies," Burns implored the crowd. "Tell them 'Hell, no!' When you make a deal with the devil, the devil is going to win in the long run."[5]

Enduring ridicule, four principal entities remained to form the new and novel Bitterroot Grizzly Bear Reintroduction Coalition: Defenders of Wildlife; the National Wildlife Federation; the Resource Organization on Timber Supply; and the Intermountain Forest Industry Association. "All of the timber and labor union guys found the dialogue we had established to be refreshing because they were used to getting verbally trashed by environmentalists," Fischer says. "We all felt good, but let's face it, the devil is in the details."

The timber industry had ample reason to be motivated. Timber volumes from public lands had declined from 12 billion board feet a year to three billion board feet during the 1990s. Feeling its back pressed against the wall, the industry also saw what had happened with government wolf reintroduction in Yellowstone and central Idaho—the American Farm Bureau Federation's attempts to derail the program were summarily pushed aside by Interior Secretary Bruce Babbitt who personally carried the first wolves into the national park, saying that "it was the right thing to do."[6]

By helping environmentalists with grizzly bear reintroduction in central Idaho, the timber industry clearly expected to receive assistance down the road with advancing timber sales that weren't considered ecologically destructive. "There is enough common ground out there that we don't need to be fighting," the late Seth Diamond of the Intermountain Forestry

Industry Association said just months before he died in an airplane crash. "We see the opportunity to create a new model for endangered species conflicts throughout the country."[7]

As they negotiated, the conservationists and the timbermen also discovered something else: they both liked to hunt in the fall, float rivers, wet fly lines for trout, and spend time with their families on camping trips. And they discovered that they actually enjoyed each other's company. The realization that they used the backcountry in similar ways proved instrumental in thwarting local politicians who were opposed to bringing back grizzlies.

In fact, in a move that stunned people like U.S. Senators Larry Craig of Idaho and Conrad Burns of Montana, who enjoyed using loggers as a fulcrum for advancing their own anti-environmental sentiments, the timber industry went to bat for the coalition and secured $250,000 from Congress for an Environmental Impact Statement, which, as Fischer observes, "Is probably the single most critical step in any species reintroduction effort."

The proposal that the coalition put together has three key provisions: (1) It would return grizzlies as an "experimental, non-essential population," a designation that provides greater flexibility for managing animals protected under the Endangered Species Act; (2) It would allow grizzlies to roam only in wilderness areas—any bears that wandered onto private land could be trapped or killed; and (3) It would give a fifteen-member locally selected "citizen management committee" influence in deciding how the bears should be stewarded. It's a policy trifecta that, interestingly enough, won a huge endorsement from then-Montana Governor Marc Racicot, a trusted confidante to George W. Bush.[8]

As Fischer says, the idea of a citizens committee isn't new, but was advocated for managing threatened species by seminal conservationist Aldo Leopold in 1936. "I am satisfied that thousands of enthusiastic conservationists would be proud of such a public trust, and many would execute it with fidelity and intelligence," Leopold wrote. "I can see in this setup more conservation than could be bought with millions of new dollars, more coordination of bureaus than Congress can get by new organization charts, more genuine contacts between factions than will ever occur in the war of the inkpots..."[9]

Needless to say, giving partial management of an endangered species to local citizens stuck in the craw of a lot of environmental organizations who believe that people in the provinces, dependent upon resource extraction, couldn't be trusted. "Weakening the Endangered Species Act for any reason should not be supported. Timber bosses and politicians may want to do that, but we support full protection of grizzlies, and strongly oppose a citizen committee with management authority," said Dr. Charles Jonkel, science advisor and founder of The Great Bear Foundation.[10]

Nevertheless, Christopher Servheen, the Fish and Wildlife Service's national Grizzly Bear Recovery Coordinator, signed off on the coalition's proposal, writing that "the re-establishment of a grizzly bear population in the Bitterroot ecosystem will increase the survival probabilities and further the conservation of the species in the lower forty-eight states. If the experimental population is lost, it will not diminish the survival probabilities for bears in other ecosystems. However, if the experimental population is successful, it will enhance grizzly bear survival and conservation over the long term by providing an additional population and thus adding a measure of security for the species." The Fish and Wildlife Service went even further. It adopted the citizens' plan as its own, infuriating Idaho's Governor, Dirk Kempthorne.

THE GOVERNOR SPEAKS

Shortly after the plan was unveiled, the Governor let fly with one of the most colorful and memorable statements ever uttered in the modern West: "Many of us have made it clear we oppose the introduction of this flesh-eating, anti-social animal into Idaho. This is probably the first federal policy that knowingly can, and will, lead to injury and death of citizens."[11]

Grizzlies, Kempthorne went on to say, were extirpated for a reason, and the reason is that humans and the predators can't get along. Former Idaho Congresswoman Helen Chenoweth-Hage added that Idahoans no more wanted grizzlies turned loose in their backyards than swimmers pined for the opportunity of having Uncle Sam dump great white sharks into the surf near their favorite beaches.

Other Idahoans weren't to be intimidated. *The Idaho State Journal*, based in Pocatello, wrote: "After last week's quotes from our state's chief executive, it's clear that Kempthorne isn't as interested in scientific research or cool analysis as he is in frightening constituents and placating the conservative lobbyists who apparently supply him with his bullet-point sheets."[12]

Amidst the debate, often inflamed by myth and superstition about grizzlies, it must be remembered that the bears do kill and eat people. However, such maulings are rare. In the Bob Marshall Wilderness, for instance, which has both hundreds of grizzlies and thousands of back-packers, horsepackers, hunters, and fishermen, just one person has died as a result of being mauled by a grizzly during the last fifty years.[13] In fact, the odds of getting killed by a grizzly in the lower forty-eight states are about the same as being killed in a commercial jet-liner crash: one in ten million. By comparison, the odds of being killed in a car accident are one in nine thousand or a thousand times greater than meeting your maker at the jaws of *Ursus arctos horribilis*.[14] Federal biologists say chances are good that the few released bears would simply disappear into their preferred life of seclusion and human avoidance, and Servheen estimates that it would take a minimum of fifty years (calculated at a 4 percent annual population growth rate) to ever reach the Fish and Wildlife Service's desired recovery target of 280 grizzlies. More likely, it would take longer, perhaps 110 years (at a 2 percent growth rate).[15]

Under the experimental, non-essential designation, ranchers could kill bears threatening their livestock, hunters could kill bears in self-defense, and wildlife managers would aggressively trap and remove bears that ambled into towns, livestock pastures, and apple orchards, particularly in the Bitterroot Valley of Montana where such ursine explorations are most likely to occur.

The coalition's plan has still another selling point: all bears would wear either radio collars or Global Positioning System transmitters, enabling their movements to be plotted in near real time. This technology would open up a whole new front in public environmental education, by allowing students in local communities to partake vicariously in the journey of bears they "adopted." Kids, outfitters, hunters, ranchers, and backpackers would know, simply by turning on their computers, when grizzlies were in the high country or in the valleys, when they had gone underground to den, and when they emerged in the spring. People would be able to look toward the mountains with knowledge and less fear.

One could thus assert that flexibility with implementing the Endangered Species Act doesn't get any more local than this, which is exactly what George W. Bush, campaigning to end the top-down management of the Clinton era, promised: "locally based decision making."

CAN CITIZENS BE TRUSTED TO DO THE RIGHT THING?

Yet, it is precisely such locally based decision making that worries people like Doug Honnold, an attorney with the Earthjustice Legal Defense Fund. "The citizens' committee is not necessarily going to be sympathetic to the needs of the bear. In the spring, when bears naturally move out of the deep snows in the mountains, following riparian corridors down to lower elevation valleys where there are people, they can be removed, no questions asked. With an 'experimental, non-essential status,' making sure they survive will not be the highest priority as it should be. Any plan should start with the science of bear biology and not put bear reintroduction into a posture of retreat and apology from the very start. If we're sincere about this, we need to figure out how we can accommodate the free-roaming behavior of bears and not let the politics of a citizens committee make that objective extremely difficult to achieve."

Consequently, Earthjustice is opposed to bringing back grizzlies under an experimental, non-essential statute. Another opponent is Mark Solomon of the Idaho Conservation League. He believes that leaving grizzly bear management in the hands of citizens who are appointed by politicians— some of whom are bitterly opposed to endangered species protection—is inherently risky. "That leaves a lot of lay people in charge of a biologically complex issue," he says, adding that if conflicts erupt, they could set back grizzly recovery and perhaps even doom bears in the lower forty-eight.

Rather than rush into reintroduction, Solomon would wait until communities surrounded by national forests make a full transition from resource extraction to being service providers for tourists and outdoor recreators. Idahoans will then be ready to see the value of the grizzly and embrace the idea that the federal government should carry out reintroduction with the full authority of the Endangered Species Act, which gives bears maximum protection.[16]

Solomon's look into the future of Idaho's economy isn't merely wishful thinking. In a landmark analysis completed by the Sonoran Institute, natural resource economist Ray Rasker notes that wildland settings, where traditional resource extraction is on the wane, boast the strongest local economies because people are drawn to the high quality of life. As to assertions that grizzlies are a wrecking ball to local economies, there is no empirical data to support this contention. In fact, the two fastest growing corners of the Northern Rockies—northwestern Montana and northwestern Wyoming—are both surrounded by public land where grizzlies roam.[17] "In both Yellowstone and the Crown of the Continent," says Tom France, "grizzlies could be seen as an economic engine, not a liability to prosperity."

But France, like others, wonders how long such a transition will take. If it took decades for a majority of residents to welcome back grizzlies, will it be too late for both the bears and for wildness? Habitat is being rapidly nibbled away by the tremendous influx of new residents. During the year 2000 alone, when the regional population of the Pacific Northwest grew by 250,000, residents paved or developed more than one hundred square miles of rural land and clear-cut seven hundred square miles of previously unlogged forest. In Montana, during the 1990s, the amount of acreage converted from ranches to land-fracturing subdivisions equaled the total area of Glacier National Park or one million acres.[18]

The Endangered Species Act is also faced with ongoing attacks from a hostile wing of the Republican party. Congressman James Hansen of Utah, for example, claims the Act was only intended to save big, popular species, not every bug, bird, turtle, or potentially man eating animal. He believes that private property owners should be offered incentives—such as payments for tolerating endangered species—rather than the Fish and Wildlife Service using the Act as a regulatory hammer to prevent development from moving forward. Similarly, he and his allies believe that the Act should not stand in the way of utilizing natural resources on public land.[19]

THE GREAT WESTERN HYPOCRISY

On a visceral level, it's easy to understand the passionate argument that governor Kempthorne makes. Reintroducing grizzly bears to Idaho is an example of detached federal wildlife managers coming in to a state they know little of and arrogantly ignoring local concerns. In the process, they create trouble where there isn't any.

But one could also point out that "the lady doth protest too much." Governor Kempthorne also knows that much of his state is federally owned and that it belongs to all American people, who have been quite magnanimous to Idahoans. Billions of dollars have been spent building logging roads into national forests so that the timber mills, which employ loggers and sawyers, can turn a profit. Citizens have also permitted ranchers to graze livestock on public lands at rates below fair market value and at a cost to native wildlife through subsidized predator control. They've shelled out handsomely to subsidize the construction of dams throughout the Columbia Basin, giving Idaho potato farmers cheap water for irrigation and the state's residents some of the cheapest electricity in the nation. They've allowed mining companies to extract hard rock minerals from the public estate without having to pay the kind of royalty generated through oil and gas drilling along the Continental Shelf. Americans have absorbed tremendous resource and service costs on behalf of the very Idahoans who now stand in the way of grizzlies, and if Americans think grizzlies deserve an opportunity to return to their former habitat in the national forests of Idaho, Kempthorne may have to acknowledge that.[20]

However, Greg Schildwachter—now a senior natural resource policy analyst in Kempthorne's office of species conservation and formerly with the Intermountain Forest Industry Association, which supports grizzly reintroduction—gave me some background on how difficult this sort of acknowledgment can be, whether it comes from the governor of Idaho or the state's people. "If you look at the case of the Bitterroot grizzly plan," Schildwachter explains, "you're brokering a situation in which nobody is getting anything that makes them completely happy. Real negotiation involves ingenuity that gets a little something for everybody."

In central Idaho, getting a little something for everybody entails all three sides—conservationists, resource extractors, and politicians who claim they are in favor of true local community collaboration—meeting at what some call "the radical center." Of course, there is an alternative to such a meeting, and that alternative is to do nothing. The result of that decision would be forests, mountains, and rivers devoid of the magnificent large animal that defined the Rockies as truly wild.

IS THIS THE END OR A BEGINNING?

The last grizzly was killed in the Bitterroot Mountains in 1932. The last confirmed grizzly track was found fourteen years later— more than half a century ago—in 1946, by a young Forest Service district ranger by the name of Bud Moore. That last bruin paw print, the final haunting piece of physical evidence that culminated millennia of grizzly existence in the Bitterroot mountains, was found the same year that the state of Idaho decided to end its grizzly sport hunting season.

We can't go back and interview people like William Henry Wright to ask him if he truly saw the end coming for bears, and if he did would he and others like him have exercised the most difficult emotion of all—self restraint. But happily, we still have Bud Moore, who today is a lucid, insightful octogenarian retired in Condon, Montana and a connection to the past. Moore knows how vacantness feels when it manifests itself in the gut, and he's remorseful about what his generation wrought.

"I didn't want to believe they [grizzlies] were gone. They had been so exciting in my youth," he says. "I didn't understand or could even spell 'ecosystem,' but I grew up out there in those woods when they were full of wildness before we hit them with dozer blade. When I never saw another sign of grizzlies again, I felt that loss. It just felt like the woods were diminished. The grizzly is about the only thing out there that is bigger in natural intellect. He's bigger than you are in damned near every way. When you run into him you've got to deal with something beyond your own self and ego."

Moore once helped his father finish off a Bitterroot grizzly that was killing sheep. Now he is trying to help Fischer, France, and loggers bring the circle of redemption back around. "There is a lot of hysteria that comes from people who are afraid," he says. "It's sincere fear based on their perspective of the world, but it's not necessarily fear that is based on reality."

His dream is that before he dies, decision makers will come together with wildlife managers and citizens "to talk and reason and even get some grudging acceptance to try bear reintroduction." He believes that towns along the Bitterroot escarpment could benefit mightily on the novelty of their accomplishment by proclaiming to tourists coming into their community: "Welcome to the Land of the Grizzly."

THE BROADER CHALLENGE

Daniel Kemmis, director of the Center for the Rocky Mountain West at the University of Montana, a former mayor of Missoula, and a guru in the local empowerment movement of "regionalism," is hopeful that something will happen to bring about just such a new chapter in wildness. Calling the architects of the Selway-Bitterroot grizzly coalition "pathfinders," Kemmis goes on to say that their plan "is the latest and one of the best in a lengthening series of western efforts to overcome our ideological differences and solve our problems here."[21]

Fischer is more succinct in his estimation of the plan to return grizzlies to the nation's second largest roadless area. "Is this a perfect solution?" he asks rhetorically. "No. I don't claim to suggest it is. It is a work in progress."

Even with its imperfections and compromises, the plan has altered people's assumptions about what is possible. It has set an example for how imperiled species could be managed at the local level in the rest of the United States, where wolves wait to be returned to the Adirondacks and the Sierra Nevada, where grizzlies, wolves, buffalo, and wapiti could reinhabit the Great Plains, where jaguars could be restored to the Huachucas and other corners of Arizona and New Mexico, and where woodland caribou and elk could once again roam the upper Midwest.

"The question now," says Kemmis, "is how do we find our way to that high, hard-won common ground?"

It won't come from playing a Sisyphean game of King of the Hill, nor from trying to shout each other down every chance we get, nor by waiting for the right stars to align over the White House.

Much of it will depend on the change in consciousness that is taking place in our culture, some of that change incremental, some of it punctuated by great leaps of understanding. William Henry Wright, the great grizzly slayer, showed us one vision of the world. With bullets and the spirit of a conqueror, he helped to win the West. Now our challenge is to win back wildness, not only in the western U.S. but also across the continent. It will take a humble reverence for embracing something more powerful and potentially more dangerous than ourselves. A few in Idaho and Montana have taken the first step.

PART IV

HEART OF THE WILD

Nature is not a place to visit, it is *home*.

— GARY SNYDER

Joined Souls

Richard Nelson

A winter morning on the Alaska coast. Kluksa Mountain heaves up through a gap in the clouds, revealing the immense, unbroken sprawl of forest on its lower slopes, the alpine meadows higher up still sheathed in winter drifts, and above that the perfect, symmetrical curve of the crater's rim. Except for a plume of snow furling off the summit, the brawl of yesterday's storm has given way to a breathless, exhausted peace.

I trek slowly along the island's shore, Keta traipsing at my side, our footsteps softened on the moist, waning snow. Distracted by a bald eagle banking and circling overhead, I'm slow to notice when Keta veers away and eagerly snuffles the surface. What she's found is a heart-shaped track, so clean and crisp, so sharp-edged, so newly wet in its basin, that I instinctively glance up, as if I might see the hoof of the deer that made it, like a pen lifting from a freshly written word. Alas, there is nothing but a field of white embossed with a single line of deer tracks, like a skein of miniature snow angels.

Keta perks her border collie ears and raises her snout into a lilt of air flooded with deer scent. She takes in the smell with as much pleasure as I'd find in a warm kitchen saturated by the aroma of baking bread. Wavering her muzzle from side to side, she steps forward almost involuntarily, drawn by the gently drifting fragrance.

Farther down the beach there's a cluster of droppings among the tracks. I squat beside Keta and push my index finger into one of the shiny, chocolate-colored lumps. It's soft, clayish, not at all sticky, there's no unpleasant odor, and amazingly I can feel in it the lingering warmth of a deer's body. In twenty years tracking black-tailed deer on this island, I've rarely found a warm scat, least of all in the winter when droppings chill within minutes.

Still crouched down, rifle in hand, I peruse the shoreline ahead: windrows of kelp left by yesterday's pounding surf, black lava boulders, jackstrawed driftwood logs, and a tangle of branches and boughs at the forest's edge. When I'm about to stand, something brown flutters among the twigs, startling into my mind the phantom image of a deer. But it's a varied thrush—apricot-breasted, a bit larger than a robin—tardy about migrating south. The bird flicks down onto the seaweed and hops around skewering beach flies numbed by the cold. I'm fascinated by varied thrushes, enthralled by their chiming voices, eager to know more about their furtive, shrouded lives; but I resist the urge to watch the hunting bird, remembering the earnest work that brings me here and mindful that a mid-winter day passes quickly.

Farther along the shore we find a different set of tracks cutting directly across the deer's line of travel. They're closely spaced, about the size of a dog's prints but with five toes instead of four. After about ten yards they suddenly transform into a sinuous, belly-dragging trough that leads toward the tide's edge. While Keta samples this new olfactory delight, the trackmaker announces itself with a series of sharp, birdlike chirps, coming not from land but from out in the slick waters of the cove. Just beyond our anchored skiff a dark shape cleaves the water. At a glance, I might have mistaken it for a harlequin duck or a green-winged teal, but those chirps are as unmistakable as the sleek back and ropy tail that break surface as the creature dives: river otter.

The animal's wake dissipates…and in the protracted calm that follows I figure that's the last I'll see of it, given the otters' mysterious ability to disappear even in large bodies of water where there's seemingly no place to hide. But to my surprise the silhouette pops up again, now oddly shaped and surging toward shore. Through binoculars, I see there's something clamped in the otter's mouth—a small flounder, fully alive, flapping spasmodically. The fish is still flapping when the otter scurries up onto a rock and begins nipping off chunks, starting at the tail and then gradually working toward the head until—at last—a fatal crunch ends it all. I've seen this same agonizing process many times and always wonder why the otter doesn't first kill its prey, a mercy that would spare the fish from watching as its own body is devoured.

However we may dream the beauty of the wild, nature gives no mandate for a quick and easy death.

At the north edge of Sea Otter Cove, the deer's tracks angle up the beach, weave through a scramble of alders, and fade back into the forest. I debate whether to follow them or stick to the shore, since it's likely other deer have come out to feed on stranded kelp, but Keta's fixation on the fresh prints is too much to resist. Judging by their size and long toe drags, these must be the marks of a buck or a hefty doe. Although some backwoods adepts claim you can distinguish a male deer's tracks from those of a female, neither my eyes nor my imagination are acute enough to see a difference.

A dense, murky dusk hovers inside the forest, the noise of surf on the outer reefs subsides to a mumble, and siskin chitterings drift down like splinters of frost. There's only a haze of snow under the towering Sitka spruce and hemlock trees, where dense boughs eclipse the sky. On the southeast Alaska coast, ancient forests like this one are a refuge for black-tailed deer in times of deep winter snow—a place where they can move around without plowing through drifts, take shelter from gales and bitter cold, browse on huckleberry and blueberry bushes, and find nutritious, ground-hugging evergreens like dwarf dogwood and five-leafed bramble.

Luckily, there's just enough snow to reveal the deer's tracks along a game trail that's worn down like a little trough in the forest floor. I motion Keta into the lead as we thread our way between the enormous, impassive trees. At intervals, I pause to check the breeze—nothing more than a faint touch of air, like a lover's whisper against my cheek, barely perceptible but laden with meaning.

In the modern world of cities where we consign ourselves to tightly housed and sheltered lives, how often do we think about the direction of a breeze? How often—apart from delicate social situations—do we consider the ineffable, wraithlike drift of our own odor? How often are we anything but oblivious to the scents given off by creatures all around us—the fragrant streamers, musky pools, and aromatic ribbons that shape so much of Keta's sensory world? Through nearly all of human history, except for our tamed and cloistered urban generations, scent and breeze have been among the paramount considerations of daily life.

By closely observing animals like dogs and deer, skilled hunters become intimately familiar with this rich but nearly undetectable dimension of their natural surroundings. For the rest of us, there is only one way to recover this lost awareness, this intricate relationship to the wind, this connection with a sense as vital as sight and hearing to many of our fellow creatures. We must take ourselves to the wild places and let them teach us once again.

The deer's tracks lead us along a game trail I have followed hundreds of times over the years—enough so that I've begun to see how the landscape changes: trees toppling, streambeds shifting, saplings growing taller, carcasses of winter-starved deer fading to scatters of bone. But now we come across the most remarkable of these changes: a string of shallow depressions, each about twelve inches long and twice the width of my boot, far enough apart so that stepping from one to another is like walking in a giant's footprints. I first came across these indentations with Keta last summer, and we followed them for almost half a mile—incomparably farther than any traces like this I'd found before. Looking closely, I noticed that each of the depressions had been scraped out, leaving a tiny midden of twigs and moss along its back edge.

I also kept an eye on the big spruce trees growing beside the trail, and eventually I found several with conspicuously worn, smoothed, almost shiny bark on the side facing the path. Some of the trunks had wounded patches where the bark had been torn off and clear sap oozed out over the exposed wood. A closer look revealed threads of umber hair and tufts of bleached brown underfur caught in the bark or clinging to the sap. One of the trees was gashed about seven feet above the ground, as if it had been sideswiped by a truck, but deep punctures and gouges showed unmistakably the work of claws and canines.

On that day last summer these signs were fresh and raw, and scent lingered thick as smoke in the forest. Keta went all awry—sniffing the pungent air, rushing ahead and turning back, peering into the thickets, torn between chase and retreat. The excitement that raged like a thunderstorm inside her body seemed to spill over into mine.

I kept thinking that if only we'd come a few hours earlier we might have encountered the maker of these signs: a hulking grizzly (or brown bear, as they're called here) working his way along the trail, scraping out signatory footprints, then lofting up on his hind legs to scrawl and gnaw and rub his back against the trunks, as if to relieve a tremendous itch. Apparently the "itch" is sexual, driven by the bear's urge to mark his territory, show his size and strength to any passing male, and advertise himself to females searching for a mate. Usually I feel secure and protected inside the forest, but then I was starkly aware that Keta and I might be considered irritating intruders or nettlesome voyeurs, and that mating season can seriously undermine the bears' normal shyness. There was also no question about this being a grizzly, since black bears do not live on this particular island.

As Keta became more and more intense, I felt increasingly puny and vulnerable. Although I've heard many stories of bears repelled by a dose of pepper spray, the canister on my belt seemed about as formidable as mosquito repellent. Much as I dislike making noise in the woods, I clattered sticks together and self-consciously, apologetically, announced myself as we tramped along.

At the same time, I was overswept by a feeling of pure elation in the presence of such feral power, elation that a creature so emblematic of wildness haunts the forest within a few miles of my front door, elation that I cannot claim sole dominance in this island's food chain, elation that those ears might take in my voice at this very moment, elation at the thought of my presence registering in the bear's mind. And elation that I can still experience the grizzly bear as a normal part of my days—this animal that has moved humans, for thousands of generations, to enter the land with humility, acknowledging that we are not necessarily in charge here.

As I walked in these tracks last summer, I was also glad to have Keta beside me, because her sharp eyes, ears, and nose make her a superb warning system, and because company of any sort gives the precious illusion of safety. But now, on this December morning, Keta is oblivious to the inert, scentless bear sign, oblivious even to a moldering pile of grizzly scat packed with berry seeds and bits of salmon bone. Somewhere on the high flank of Kluksa Mountain, in a cavern underneath the snow, the hibernating bear sighs long, languid breaths. And here beside me, fixed in the moment, Keta's lively mind cares only about the invisible but immediately present deer.

The hoofprints lead us away from shore and up a long, steep slope where the trail is churned to dark earth by constant deer traffic. I pause to catch my breath beneath a huge Sitka spruce that bows out over a cliff, its outsized roots braced against the hillside below and clasping the stone like pythons. The trunk itself bends up to vertical and reaches almost two hundred feet into the canopy.

As a child, I was taught that trees have no capacity to think, but this venerable spruce, no less alive than I am, is clearly responding to its circumstances, just as I catch my balance if I'm falling. In the objective language of science, we can explain the marvelous way this tree shapes itself as a concerted physical reaction by millions of cells within its body; but we can also explain human thought as a concerted physical reaction by millions of our own nerve cells. I'm comfortable believing that trees think, but they do it very slowly. In the several hundred years this spruce has held its place here, I am certain it has come to know—in ways utterly beyond my grasp—the surrounding forest, the turn of winter and summer, the trees with which its roots are intertwined, the earth and rocks on which it grows, and perhaps even those of us who pass by as fleetingly as cloud shadows.

I spent some of the most important years of my life as an anthropologist living with Koyukon Indians, traditional hunter-gatherers whose log cabin villages stand beside the Yukon and Koyukuk Rivers in the northern forest of Alaska. Koyukon elders told me that trees—like all other plants, like every kind of animal, and like the land itself—are gifted with a consciousness similar to our own, as well as a potent and attentive spiritual power. In the Koyukon way, a spruce tree is aware of the camper sleeping under its boughs, and the tree protects that person from spiritual or physical danger. The spruce also willingly gives itself to someone who cuts it down, but in return that person must show respect for the tree. Disobeying one of the rules for proper behavior toward trees can bring bad luck to the violator, who might then have difficulty finding trees of good quality.

I'll never fully comprehend this way of seeing trees, and I cannot share the Koyukon people's certainty that our world is filled with spirit, power, and awareness; but I believe there is an unimpeachable wisdom in this way of living—wisdom in treating all of nature as if it were permeated with sacredness, in behaving as if our actions toward the earth and every living thing were judged by powers greater than ourselves, and in acknowledging that food and water and air come to us as gifts deserving our profound gratitude.

Where the deer's tracks reach the top of the slope, I slip behind a con-cealing tree and look out over a mounded terrace cloaked in deep, cushiony sphagnum, mostly snow-covered at this higher elevation, and overshad-owed by a cathedral forest of immense, columnar trees. I've often spotted deer in this open parklike woods, but nothing moves, and all is silent except for a downy woodpecker spearing dormant insects from crevices in the bark.

Scattered through the forest are rotted stumps and fallen tree trunks called "nurse logs," each with an orderly row of saplings along the top, like birds on a wire, their tiny rootlets probing the heart of the decaying log for nourishment—new life arising from old. I'm reminded that exactly the same process unfolds every summer in Bear Creek, less than a mile up the coast from here. When I trekked along the creek last August, thousands of spawning pink and chum salmon circled in the deep pools, crowded the eddies, and skittered across the riffles—all of them destined to die and decay within a few weeks, fertilizing the waters with their own bodies, re-entering the food chain that will nourish their own minnowy offspring. In this way, each generation of salmon feeds the next. The forest is also nour-ished by salmon, when river otters, bald eagles, and grizzly bears leave fish carcasses and their own droppings on the forest floor. In this way, for countless thousands of years, gifts from the far ocean have flowed into the stream, into the soil, and into the veins of trees.

I, too, have salmon in my veins, and in this regard I share a common source with salmon fingerlings in the creek and with ancient trees in the forest. Even the deer we follow are nourished by salmon through the plants growing in this soil. Each of us participates in the same living process, pass-ing life back and forth between our generations. For me, these interconnections aren't simply abstractions or turns of language. They are the tangible realities underlying my own existence...on this day, tracking this deer, on this island from which a part of my life is drawn. My inten-tions—no different from the bear's in his season—are those of an omnivorous predator. I have come here to fill my veins.

Slats of sky blink brightly in the spaces between trees as we approach a stretch of open ground. Noticing that Keta's picked up another strong eddy of scent, I step slowly and quietly, holding still for long minutes, watching as keenly as I know how, as we enter a muskeg—a broad, boggy meadow scattered with dwarfish shore pines and scrubby junipers, blanketed by a few inches of snow. Tall spruce, hemlock, and cedar trees crowd tightly around the muskeg's edge and climb a steep ridge at the far end. Looming beyond the ridge is the prodigious, inward-curving wall of Kluksa Mountain, draped with lacy tendrils of fog and glimmering in the low, heatless sun.

I spend a few minutes taking it all in, feeling almost embarrassed by the blind luck to be here, pitying the millionaires and princes whose riches cannot buy the luxury of this freedom, lavishing myself in a place as wild and pristine as any on earth, toasting the paradise of home, feeling grateful yet again for the privilege of living in a nation where I can wander such a land as this at will.

The island where I stand is part of the Tongass National Forest, which comprises most of the surrounding Alexander Archipelago and the adjoining Alaska mainland. It is America's biggest and most spectacular national forest, an area the size of West Virginia, embracing dozens of large islands and over several thousand smaller ones; a landscape of inlets and fjords, rivers and glaciers, rugged mountains and plunging valleys; containing the largest temperate rainforest left on earth, with an interlacing mosaic of muskegs, meadows, estuaries, thickets, and high tundra; inhabited by grizzly bear, black bear, wolverine, gray wolf, mountain goat, black-tailed deer, pine marten, beaver, river otter, sea otter, raven, bald eagle, peregrine falcon, hermit thrush, northern goshawk, black oystercatcher, harlequin duck, the river-thronging salmon, and hundreds of other creatures large or small, abundant or rare, ordinary or charismatic; a place where not one species of plant or animal has vanished in historic memory; the wild earth as it was meant to be—the Great Raven's world in its full and lavish diversity.

The Tongass National Forest belongs to an assortment of public wildlands all across this country, including many that rank among the world's most remarkable natural places. Protecting these national treasures and keeping them open to everyone stands with the most brilliant achievements of American democracy, a transcendent legacy to humankind, equal to our highest intellectual, artistic, social, political, and technological accomplishments. Public wildlands are significant not because of what we have created here, but because we have deliberately chosen not to create anything so as to preserve the original masterwork of creation itself.

These lands are also a powerful expression of democratic freedom—places where we are free to enter and explore, free to experience peace and beauty and solitude, free to witness the extravagant living community of which we are a part, free to braid ourselves into the earth on which this nation stands. Unlike so many lands held privately, our public wildlands are not locked up—they are locked open.

Exploring these lands—from the Tongass to the Everglades, Yellowstone to Yosemite, Cape Cod to Point Reyes—creates in me a deep sense of patriotic allegiance toward the American land, embodied by these places of which I am a stakeholder. It has also led me to an ethic of responsibility toward these lands, supporting the people and groups that work to protect them. As a conservation activist, I recognize the privilege of living in a democracy where I can speak my opinion and advocate for responsible use of our lands. Each star on the American flag reminds me, with pride and gratitude, of the land within every state that is open to me by right, heritage, and citizenship.

I never feel more fortunate, more free, and more alive than when I am afoot in the wild country of home, and above all when I am on this island where my soul resides. I suspect Keta feels much the same, judging by her excitement every time we come here. But right now her wet nose on my hand signals that she's impatient to move along.

Just after we entered this muskeg, the tracks we had followed for almost half a mile switched abruptly from a walk to a run, cut away from the trail, and disappeared into a frazzle of underbrush. Keta was dismayed when I chose to abandon them, but almost certainly the deer had heard us coming, watched until we showed ourselves, then headed for cover. I'm surprised we didn't see it, but even when a deer stands in plain sight it can seem invisible, as if there were only the breath of a deer drifting on the wind. Koyukon hunters told me that animals can vanish purely by intention, or they can choose to become visible and "give themselves" to a hunter—especially one who talks respectfully about them, does not brag about hunting, takes nothing beyond need, and uses each animal in the proper ways.

A raven brings all this to mind, as it wheels above us pouring out sonorous croaks and chortles, black as a shard from the midnight sky, eyes glinting, wings whistling. And once again I wonder what he could be thinking. In Koyukon tradition, Raven is the creator of our world—a feathered ebony god and a comedic genius, both powerful and mischievous. Grandfather Raven's exploits are recounted in sacred stories passed on by the elders: stories detailing how the world came to be and showing how to live respectfully, harmoniously, and responsibly in the environment to which we all belong. Today, Raven's descendents, like the one pirouetting overhead, keep a benevolent and perhaps bemused watch over humankind.

Koyukon hunters advise that a raven's flight can show the way toward game, so I pay close attention as the bird dwindles off along the muskeg's edge, lets out an echoing yodel, then swings north and vanishes beyond the ridge. Keta stares in the same direction and steps eagerly into a breeze carrying new scent. Who am I to question these two authorities? So I follow Keta, snow muffling my steps, keeping a close eye on the forest's tanglework edge. I step...pause...listen...watch...step again...pause. Shivering branches reveal the hidden breeze. A loose mob of kinglets and chickadees flits from tree to tree, clinging acrobatically to limber twigs, probing brittle cones, slivering off bits of bark.

A hundred yards farther into the muskeg fresh deer tracks crosscut our path. Keta goes on full alert, ears perked, muscles tight, nostrils twitching and flaring, as we plunge into a dense meander of scent. When I bend down to stroke her, she feels so electrified it's a wonder that sparks don't flash from her silky black fur. I move slowly, meticulously, and so quietly that my own breathing sounds like storm gusts in my ears. Sunshine breaks out through diminishing clouds; the snow is a blinding crystal sea; the shape of every twig and needle carves itself into the clear vitreous of my eye; the black crucifix of a raven soars across the blinding white of Kluksa Mountain's face.

Keta halts, locks her eyes on a clump of shore pines straight ahead, slumps down halfway to the snow, and leans forward as if she's ready to burst away but her nerves are frozen. I stop instantly, trace the line of her gaze, see nothing...and nothing...and nothing yet again, although I know without question that Keta has found a deer. Words I once heard from a Yup'ik Eskimo woman run through my mind: "If you keep still, you see what moves."

At last I pick out a twitch in the muskeg's fringe—revealing the thick, brown body of a buck, head and antlers and neck clearly visible, chest and belly and haunch mostly hidden behind a veil of boughs, a wide V of ears cupped straight toward us. I sink down behind a solitary, slender pine, snow quickly wetting my knees. The breeze surges and fades. The mid-winter sun feels faintly warm on my shoulders. The raven's voice tremors somewhere nearby.

I raise the rifle, rest it against the tree, and align the sights with their mark, precisely at the moment when the deer flicks his tail—revealing his nervous intention to move.

I will hunt a deer only if it is very close and standing perfectly still, like an offering. So I exhale and wait. Within seconds the buck wags his tail again, pulls back his ears, turns his face toward the trees, and whisks away. My arms and hands loosen, I lean my head aside and lower the rifle, listen to the last hoofbeats, then pat Keta to quiet her down. She's beyond consoling, her body a rage of unconsummated energy, her eyes begging me to follow the deer—but there's no hope of seeing it again.

When we find a sheltered place for lunch, the noon sun is barely ten degrees above the horizon and it's only a few hours until evening. Keta snuggles against my leg, eager for bits of smoked fish and venison jerky. Deer season is almost over and I'm anxious to bring in the year's venison supply, plus some extra to share with our nonhunting friends. Deer and salmon are daily staples in our house—healthy organic foods that come to us through our own labor, foods that bind my partner, Nita, and me together in the work of preparation, foods that take us into the wild and bring the wild into us, creating a communion of body and spirit with our place on earth.

Even when I take a drink from the frigid, shimmering creek beside our picnic spot, I feel myself entering more deeply into the island's life. I come here not just as a sightseer, not just as a lover of the wild, but also as an organism questing for sustenance—the food and water that I am made from—and so my sense of belonging here is ecological as much as it is emotional, aesthetic, and spiritual.

This is true not just because I live at the wild edge in Alaska, but because of the ways I have chosen to integrate with my home environment. When I visit other states—California, Wisconsin, Texas, Montana, Florida,

New York—I'm amazed by the diversity and abundance of wild game and fish: deer, elk, black bear, rabbits, ducks, geese, salmon, bass, catfish; and by the tremendous variety of edible wild plants. Compared to Alaska, it would be far easier to harvest food from the land in the warmer, more temperate latitudes of this continent. This is true even within close reach of very large cities. Anywhere on this continent, it is possible to nourish not only the soul but also the body from the wild at hand.

Two hours later, the sun is low and there's a chill in the shady places. We've searched widely in the muskeg and forest but haven't seen another deer, so I finally relent and turn back toward shore, following our earlier tracks on the crusting snow. I'm still moving slowly, still concentrating almost hypnotically on the labyrinth of trees ahead, yet I've lapsed into a momentary daydream. Then a sudden, sharp scuttling and a flurry of metallic chatters sets my heart pounding. A red squirrel skitters up a behemoth spruce almost within touching distance, clamps onto a branch, and unleashes another piercing complaint, tail twitching, eyes aglitter, chest pumping—as if a Fourth of July sparkler had come to life. I worry that the racket could alert deer to my presence, but these hyperactive Chicken Littles throw so many tantrums that neighboring animals probably ignore them.

Moments later we step out onto the beach: lava bedrock patched with gleaming snow, kelp ribbons embroidered with frost, and the limitless sky aflame. A litter of seagull feathers lies just above the tide's reach, the surrounding snow is scuffled with talon prints, and a bald eagle stares down from a treetop. I've seen eagles pursue gulls dozens of times, watched the mirrored flight of predator and prey, the slow motion dodging and jinking, the two birds often just inches apart. And feather piles like these are a common find, but I have yet to witness the kill. I can only imagine the captured gull flapping helplessly, the golden beak bending down, and the mingling of flesh and blood that follows, as gull becomes eagle.

As for me, the hunting seems finished, but this day has been an immersion in pure winter glory, a resounding success by any measure. Keta stays close as we turn north toward the cove and the waiting skiff. But something impels me to stop and glance behind...and near the place where we emerged from the forest, a doe steps down the shore, crossing the tracks we've just made. Our scent must lay in there like fog, but she shows no alarm. Nor, as far as I can tell, does she notice Keta and me standing in plain sight, hardly more than whispering distance away.

Then, before my thoughts catch up with my senses, a dark, bulky shape materializes where the doe came out moments before—a buck deer, head low, plunging his muzzle into air rich with female fragrance. His beamy antlers glow burnished red in the late sun. He moves quickly, leans forward, and touches his snout to the doe's tail as she steps down off the snow onto cobbles left wet and black by the receding tide. His mouth is partly open, his upper lip curled; I can see the white of his eye as he strains to look at the doe directly in front of him. He is driven wholly by desire, his normal stealth and wariness abandoned to the fervor of the rut.

I stand there for a moment, stunned into immobility. Then the doe either sees me or realizes she's blundered into the pall of our scent. She flinches sharply, turns back up toward the woods, and looks our way, her body stiff and tense and enervated, like a sleepwalker awakening in the middle of a freeway. But instead of dashing into the thicket she freezes, hard as cold iron on the ivory pane of snow.

The buck, still down on the bare, tide-washed rocks, detects her fear, as a person might feel static in the air before lightning strikes. For a moment, he shakes free of his breeding daze and looks first toward me, then at her, still uncertain which is more important.

Meanwhile, I've sat down awkwardly on the hard, snow-covered stones. During these confused moments, while the deer waver between fear and ambivalence, while they stare across the dazzle of snow and sunset and violet-blue sky, I lift the rifle to my shoulder, brace my elbows on my knees, and steady the sights just below the antlers—the one place certain to bring an instantaneous, painless end. But with the stones poking into me, I feel contorted and unsteady, and I cannot absolutely trust my aim.

During these moments, it's as if the three of us are caught in a tightening web, like insects snared in spider silk. I splay slowly down, lie prone on the cold snow, then lift the rifle again. Breaking her pose, the doe struts hesitantly toward the woods.

The buck looks after her but makes no move...as if movement is beyond him, as if he has chosen this place on earth, this winter afternoon, this instant. As if something of himself had already moved inside me and something of myself were carried into him. As if he had seen his own eyes reflecting in mine, and had willed himself to that. As if we've come here to resolve the ambiguities that lie at the crux of life.

I am carried away on the current of his willingness, wholly lost in the world outside myself, yet entirely, perfectly within. My body is filled with all the passion and anguish and ecstasy of a love that stays forever in the heart but always beyond reach, both consummated and unrequited, and destined to remain so forever.

There, suspended between conflicting imperatives, lying on a bank of crystal rain, at the seam of ocean and land, in the silent and waiting stare of gulls, with the boreal wind breathing in my ears, I release a heart-wrenching yet mindful power.

And in the sear and thunder, the shock and blindness—the fluttering flag-fall of the buck and the fading hoofbeats of the doe—there is no boundary between the beginning and the end. I see myself again, born of a beloved animal.

The surge runs over gleaming rocks and washes back against the buck's flanks. I stumble to where he lies, grasp his forelegs, pull him up onto the snow, and touch the softness of his eye to be sure that all is stilled. Then I say my thanks, while a raven calls above the shore.

I wonder if any human can learn vicariously what the hunter knows, what pours through him like a waterfall when an animal gives him life. Perhaps a farmer knows it, when his scythe cuts through tender green veins and the living wheat falls. Perhaps a gardener, when she pulls the rooted plant tenderly from moist soil. Perhaps a fisherman, when he places his shining catch on the riverbank. Perhaps the same knowledge belongs to all who watch their own hands doing the work that transforms one life into another. Perhaps everyone who participates fully in the process of their sustenance understands in the same way and sees how each of us—wherever and however we live—are made from the soil and the waters, made from the plants and the animals, made from all that is beautiful and wild in the earth itself.

Night is leaning down by the time we're in the boat crossing Haida Strait toward home. Up ahead, a path of light from the early rising moon quavers on restless water. The skiff rises on a gentle swell, drops slowly into the trough, then lifts again on the ocean's heaving chest. Behind us,

Kluksa Mountain's summit looms like a ghost against the prodigious depths of sky. The island stretches across our wake, broadening and darkening, becoming dreamlike in the distance, and radiant with the abiding, illimitable power of wildness.

Through the icy winter breeze, I feel a sudden rush of heat against my face. My breath. A deer's breath.

And I whisper thanks again.

Appendix

WILDERNESS ACT OF 1964

Public Law 88-577 (16 U.S. C. 1131-1136)
88th Congress, Second Session
September 3, 1964

AN ACT

To establish a National Wilderness Preservation System for the permanent good of the whole people, and for other purposes.
Be it enacted by the Senate and House of Representatives of the United States of America in Congress assembled,

SHORT TITLE

SEC. 1. This Act may be cited as the "Wilderness Act".

WILDERNESS SYSTEM ESTABLISHED STATEMENT OF POLICY

SEC. 2. (a) In order to assure that an increasing population, accompanied by expanding settlement and growing mechanization, does not occupy and modify all areas within the United States and its possessions, leaving no lands designated for preservation and protection in their natural condition, it is hereby declared to be the policy of the Congress to secure for the American people of present and future generations the benefits of an enduring resource of wilderness. For this purpose there is hereby established a National Wilderness Preservation System to be composed of federally owned areas designated by Congress as "wilderness areas", and these shall be administered for the use and enjoyment of the American people in such manner as will leave them unimpaired for future use and enjoyment as wilderness, and so as to provide for the protection of these areas, the preservation of their wilderness character, and for the gathering and dissemination of information regarding their use and enjoyment as wilderness; and no Federal lands shall be designated as "wilderness areas" except as provided for in this Act or by a subsequent Act.

(b) The inclusion of an area in the National Wilderness Preservation System notwithstanding, the area shall continue to be managed by the Department and agency having jurisdiction thereover immediately before its inclusion in the National Wilderness Preservation System unless otherwise provided by Act of Congress. No appropriation shall be available for the payment of expenses or salaries for the administration of the National Wilderness Preservation System as a separate unit nor shall any appropriations be available for additional personnel stated as being required solely for the purpose of managing or administering areas solely because they are included within the National Wilderness Preservation System.

DEFINITION OF WILDERNESS

(c) A wilderness, in contrast with those areas where man and his own works dominate the landscape, is hereby recognized as an area where the earth and its community of life are untrammeled by man, where man himself is a visitor who does not remain. An area of wilderness is further defined to mean in this Act an area of undeveloped Federal land retaining its primeval character and influence, without permanent improvements or human habitation, which is protected and managed so as to preserve its natural conditions and which (1) generally appears to have been affected primarily by the forces of nature, with the imprint of man's work substantially unnoticeable; (2) has outstanding opportunities for solitude or a primitive and unconfined type of recreation; (3) has at least five thousand acres of land or is of sufficient size as to make practicable its preservation and use in an unimpaired condition; and (4) may also contain ecological, geological, or other features of scientific, educational, scenic, or historical value.

NATIONAL WILDERNESS PRESERVATION SYSTEM—
EXTENT OF SYSTEM

SEC. 3. (a) All areas within the national forests classified at least 30 days before the effective date of this Act by the Secretary of Agriculture or the Chief of the Forest Service as "wilderness", "wild", or "canoe" are hereby designated as wilderness areas. The Secretary of Agriculture shall-

(1) Within one year after the effective date of this Act, file a map and legal description of each wilderness area with the Interior and Insular Affairs Committees of the United States Senate and the House of Representatives, and such descriptions shall have the same force and effect as if included in this Act: Provided, however, That correction of clerical and typographical errors in such legal descriptions and maps may be made.

(2) Maintain, available to the public, records pertaining to said wilderness areas, including maps and legal descriptions, copies of regulations governing them, copies of public notices of, and reports submitted to Congress regarding pending additions, eliminations, or modifications. Maps, legal descriptions, and regulations pertaining to wilderness areas within their respective jurisdictions also shall be available to the public in the offices of regional foresters, national forest supervisors, and forest rangers.

(b) The Secretary of Agriculture shall, within ten years after the enactment of this Act, review, as to its suitability or nonsuitability for preservation as wilderness, each area in the national forests classified on the effective date of this Act by the Secretary of Agriculture or the Chief of the Forest Service as "primitive" and report his findings to the President. The President shall advise the United States Senate and House of Representatives of his recommendations with respect to the designation as "wilderness" or other reclassification of each area on which review has been completed, together with maps and a definition of boundaries. Such advice shall be given with respect to not less than one-third of all the areas now classified as "primitive" within three years after the enactment of this Act, not less than two-thirds within seven years after the enactment of this Act, and the remaining areas within ten years after the enactment of this Act. Each recommendation of the President for designation as "wilderness" shall become effective only if so provided by an Act of Congress. Areas classified as "primitive" on the effective date of this Act shall continue to be administered under the rules and regulations affecting such areas on the effective date of this Act until Congress has determined otherwise. Any such area may be increased in size by the President at the time he submits his recommendations to the Congress by not more than five thousand acres with no more than one thousand two hundred and eighty acres of such increase in any one compact unit; if it is proposed to increase the size of any such area by more than five thousand acres or by more than one thousand two hundred and eighty acres in any one compact unit the increase in size shall not become effective until acted upon by Congress. Nothing herein contained shall limit the President in proposing, as part of his recommendations to Congress, the alteration of existing boundaries of primitive areas or recommending the addition of any contiguous area of national forest lands predominantly of wilderness value. Notwithstanding any other provisions of this Act, the Secretary of Agriculture may complete his review and delete such area as may be necessary, but not to exceed seven thousand acres, from the southern tip of the Gore Range-Eagles Nest Primitive Area, Colorado, if the Secretary determines that such action is in the public interest.

(c) Within ten years after the effective date of this Act the Secretary of the Interior shall review every roadless area of five thousand contiguous acres or more in the national parks, monuments and other units of the national park system and every such area of, and every roadless island within, the national wildlife refuges and game ranges, under his jurisdiction on the effective date of this Act and shall report to the President his recommendation as to the suitability or nonsuitability of each such area or island for preservation as wilderness. The President shall advise the President of the Senate and the Speaker of the House of Representatives of his recommendation with respect to the designation as wilderness of each such area or island on which review has been completed, together with a map thereof and a definition of its boundaries. Such advice shall be given with respect to not less than one-third of the areas and islands to be reviewed under this subsection within three years after enactment of this Act, not less than two-thirds within seven years of enactment of this Act, and the remainder within ten years of enactment of this Act. A recommendation of the President for designation as wilderness shall become effective only if so provided by an Act of Congress. Nothing contained herein shall, by implication or otherwise, be construed to lessen the present statutory authority of the Secretary of the Interior with respect to the maintenance of roadless areas within units of the national park system.

(d) (1) The Secretary of Agriculture and the Secretary of the Interior shall, prior to submitting any recommendations to the President with respect to the suitability of any area for preservation as wilderness-

(A) give such public notice of the proposed action as they deem appropriate, including publication in the Federal Register and in a newspaper having general circulation in the area or areas in the vicinity of the affected land;

(B) hold a public hearing or hearings at a location or locations convenient to the area affected. The hearings shall be announced through such means as the respective Secretaries involved deem appropriate, including notices in the Federal Register and in newspapers of general circulation in the area: Provided, That if the lands involved are located in more than one State, at least one hearing shall be held in each State in which a portion of the land lies;

(C) at least thirty days before the date of a hearing advise the Governor of each State and the governing board of each county, or in Alaska the borough, in which the lands are located, and Federal departments and agencies concerned, and invite such officials and Federal agencies to submit their views on the proposed action at the hearing or by no later than thirty days following the date of the hearing.

(2) Any views submitted to the appropriate Secretary under the provisions of (1) of this subsection with respect to any area shall be included with any recommendations to the President and to Congress with respect to such area.

(e) Any modification or adjustment of boundaries of any wilderness area shall be recommended by the appropriate Secretary after public notice of such proposal and public hearing or hearings as provided in subsection (d) of this section. The proposed modification or adjustment shall then be recommended with map and description thereof to the President. The

President shall advise the United States Senate and the House of Representatives of his recommendations with respect to such modification or adjustment and such recommendations shall become effective only in the same manner as provided for in subsections (b) and (c) of this section.

USE OF WILDERNESS AREAS

SEC.4. (a) The purposes of this Act are hereby declared to be within and supplemental to the purposes for which national forests and units of the national park and national wildlife refuge systems are established and administered and-

(1) Nothing in this Act shall be deemed to be in interference with the purpose for which national forests are established as set forth in the Act of June 4, 1897 (30 Stat. 11), and the Multiple Use Sustained-Yield Act of June 12, 1960 (74 Stat. 215).

(2) Nothing in this Act shall modify the restrictions and provisions of the Shipstead-Nolan Act (Public Law 539, Seventy first Congress, July 10, 1930; 46 Stat. 1020), the Thye-Blatnik Act (Public Law 733, Eightieth Congress, June 22, 1948; 62 Stat. 568), and the Humphrey-Thye-Blatnik-Andresen Act (Public Law 607, Eighty-fourth Congress, June 22, 1956; 70 Stat. 326), as applying to the Superior National Forest or the regulations of the Secretary of Agriculture.

(3) Nothing in this Act shall modify the statutory authority under which units of the national park system are created. Further, the designation of any area of any park, monument, or other unit of the national park system as a wilderness area pursuant to this Act shall in no manner lower the standards evolved for the use and preservation of such park, monument, or other unit of the national park system in accordance with the Act of August 25, 1916, the statutory authority under which the area was created, or any other Act of Congress which might pertain to or affect such area, including, but not limited to, the Act of June 8, 1906 (34 Stat. 225; 16 U.S.C. 432 et seq.); section 3(2) of the Federal Power Act (16 U.S.C. 796(2)): and the Act of August 21, 1935 (49 Stat. 666; 16 U.S.C. 461 et seq.).

(b) Except as otherwise provided in this Act, each agency administering any area designated as wilderness shall be responsible for preserving the wilderness character of the area and shall so administer such area for such other purposes for which it may have been established as also to preserve its wilderness character. Except as otherwise provided in this Act, wilderness areas shall be devoted to the public purposes of recreational, scenic, scientific, educational, conservation, and historical use.

PROHIBITION OF CERTAIN USES

(c) Except as specifically provided for in this Act, and subject to existing private rights, there shall be no commercial enterprise and no permanent road within any wilderness area designated by this Act and, except as necessary to meet minimum requirements for the administration of the area for the purpose of this Act (including measures required in emergencies involving the health and safety of persons within the area), there shall be no temporary road, no use of motor vehicles, motorized equipment or motorboats, no landing of aircraft, no other form of mechanical transport, and no structure or installation within any such area.

SPECIAL PROVISIONS

(d) The following special provisions are hereby made:

(1) Within wilderness areas designated by this Act the use of aircraft or motorboats, where these uses have already become established, may be permitted to continue subject to such restrictions as the Secretary of Agriculture deems desirable. In addition, such measures may be taken as may be necessary in the control of fire, insects, and diseases, subject to such conditions as the Secretary deems desirable.

(2) Nothing in this Act shall prevent within national forest wilderness areas any activity, including prospecting, for the purpose of gathering information about mineral or other resources, if such activity is carried on in a manner compatible with the preservation of the wilderness environment. Furthermore, in accordance with such program as the Secretary of the Interior shall develop and conduct in consultation with the Secretary of Agriculture, such areas shall be surveyed on a planned, recurring basis

consistent with the concept of wilderness preservation by the Geological Survey and the Bureau of Mines to determine the mineral values, if any, that may be present; and the results of such surveys shall be made available to the public and submitted to the President and Congress.

(3) Notwithstanding any other provisions of this Act, until midnight December 31, 1983, the United States mining laws and all laws pertaining to mineral leasing shall, to the same extent as applicable prior to the effective date of this Act, extend to those national forest lands designated by this Act as "wilderness areas"; subject, however, to such reasonable regulations governing ingress and egress as may be prescribed by the Secretary of Agriculture consistent with the use of the land for mineral location and development and exploration, drilling, and production, and use of land for transmission lines, waterlines, telephone lines, or facilities necessary in exploring, drilling, producing, mining, and processing operations, including where essential, the use of mechanized ground or air equipment and restoration as near as practicable of the surface of the land disturbed in performing prospecting, location, and, in oil and gas leasing, discovery work, exploration, drilling, and production, as soon as they have served their purpose. Mining locations lying within the boundaries of said wilderness areas shall be held and used solely for mining or processing operations and uses reasonably incident thereto; and hereafter, subject to valid existing rights, all patents issued under the mining laws of the United States affecting national forest lands designated by this Act as wilderness areas shall convey title to the mineral deposits within the claim, together with the right to cut and use so much of the mature timber therefrom as may be needed in the extraction, removal, and beneficiation of the mineral deposits, if needed timber is not otherwise reasonably available, and if the timber is cut under sound principles of forest management as defined by the national forest rules and regulations, but each such patent shall reserve to the United States all title in or to the surface of the lands and products thereof, and no use of the surface of the claim or the resources therefrom not reasonably required for carrying on mining or prospecting shall be allowed except as otherwise expressly provided in this Act: Provided, That, unless hereafter specifically authorized, no patent within wilderness areas designated by this Act shall issue after December 31, 1983, except for the

valid claims existing on or before December 31, 1983. Mining claims located after the effective date of this Act within the boundaries of wilderness areas designated by this Act shall create no rights in excess of those rights which may be patented under the provisions of this subsection. Mineral leases, permits, and licenses covering lands within national forest wilderness areas designated by this Act shall contain such reasonable stipulations as may be prescribed by the Secretary of Agriculture for the protection of the wilderness character of the land consistent with the use of the land for the purposes for which they are leased, permitted, or licensed. Subject to valid rights then existing, effective January 1, 1984, the minerals in lands designated by this Act as wilderness areas are withdrawn from all forms of appropriation under the mining laws and from disposition under all laws pertaining to mineral leasing and all amendments thereto.

(4) Within wilderness areas in the national forests designated by this Act, (1) the President may, within a specific area and in accordance with such regulations as he may deem desirable, authorize prospecting for water resources, the establishment and maintenance of reservoirs, water-conservation works, power projects, transmission lines, and other facilities needed in the public interest, including the road construction and maintenance essential to development and use thereof, upon his determination that such use or uses in the specific area will better serve the interests of the United States and the people thereof than will its denial; and (2) the grazing of livestock, where established prior to the effective date of this Act, shall be permitted to continue subject to such reasonable regulations as are deemed necessary by the secretary of Agriculture.

(5) Other provisions of this Act to the contrary notwithstanding, the management of the Boundary Waters Canoe Area, formerly designated as the Superior, Little Indian Sioux, and Caribou Roadless Areas, in the Superior National Forest. Minnesota, shall be in accordance with regulations established by the Secretary of Agriculture in accordance with the general purpose of maintaining, without unnecessary restrictions on other uses, including that of timber, the primitive character of the area, particularly in the vicinity of lakes, streams, and portages: Provided, That nothing in this Act shall preclude the continuance within the area of any already

established use of motorboats.

(6) Commercial services may be performed within the wilderness areas designated by this Act to the extent necessary for activities which are proper for realizing the recreational or other wilderness purposes of the areas.

(7) Nothing in this Act shall constitute an express or implied claim or denial on the part of the Federal Government as to exemption from State water laws.

(8) Nothing in this Act shall be construed as affecting the jurisdiction or responsibilities of the several States with respect to wildlife and fish in the national forests.

STATE AND PRIVATE LANDS WITHIN WILDERNESS AREAS

SEC. 5. (a) In any case where State-owned or privately owned land is completely surrounded by national forest lands within areas designated by this act as wilderness, such State or private owner shall be given such rights as may be necessary to assure adequate access to such State-owned or privately owned land by such State or private owner and their successors in interest, or the State-owned land or privately owned land shall be exchanged for federally owned land in the same State of approximately equal value under authorities available to the Secretary of Agriculture: Provided, however, That the United States shall not transfer to a State or private owner any mineral interests unless the State or private owner relinquishes or causes to be relinquished to the United States the mineral interest in the surrounded land.

(b) In any case where valid mining claims or other valid occupancies are wholly within a designated national forest wilderness area, the Secretary of Agriculture shall, by reasonable regulations consistent with the preservation of the area as wilderness, permit ingress and egress to such surrounded areas by means which have been or are being customarily enjoyed with respect to other such areas similarly situated.

(c) Subject to the appropriation of funds by Congress, the Secretary of Agriculture is authorized to acquire privately owned land within the perimeter of any area designated by this Act as wilderness if (1) the owner concurs in such acquisition or (2) the acquisition is specifically authorized by Congress.

GIFTS, BEQUESTS, AND CONTRIBUTIONS

SEC. 6. (a) The Secretary of Agriculture may accept gifts or bequests of land within wilderness areas designated by this Act for preservation as wilderness. The Secretary of Agriculture may also accept gifts or bequests of land adjacent to wilderness areas designated by this Act for preservation as wilderness if he has given sixty days advance notice thereof to the President of the Senate and the Speaker of the House of Representatives. Land accepted by the Secretary of Agriculture under this section shall become part of the wilderness area involved. Regulations with regard to any such land may be in accordance with such agreements, consistent with the policy of this Act, as are made at the time of such gift, or such conditions, consistent with such policy, as may be included in, and accepted with, such bequest.

(b) The Secretary of Agriculture or the Secretary of the Interior is authorized to accept private contributions and gifts to be used to further the purposes of this Act.

ANNUAL REPORTS

SEC. 7. At the opening of each session of Congress, the Secretaries of Agriculture and Interior shall jointly report to the President for transmission to Congress on the status of the wilderness system, including a list and descriptions of the areas in the system, regulations in effect, and other pertinent information, together with any recommendations they may care to make.

Approved September 3, 1964.

LEGISLATIVE HISTORY:

House Reports:

No. 1538 accompanying H.R. 9070 (Committee on Interior & Insular Affairs)

No. 1829 (Committee of Conference).

Senate Report:

No. 109 (Committee on Interior & Insular Affairs).

Congressional Record:

Vol. 109 (1963):

April 4, 8, considered in Senate.

April 9, considered and passed Senate.

Vol. 110 (1964):

July 28, considered in House.

July 30, considered and passed House, amended, in lieu of H.R. 9070.

August 20, House and Senate agreed to conference report.

Notes

INTRODUCTION ~ TED KERASOTE

1. *World Resources 2000-2001*, United Nations Development Programme, United Nations Environment Programme, World Bank, World Resources Institute, Amsterdam, 2000, p. 24.

2. World Commission on Protected Areas website: *http://wcpa.iucn.org/wpc/wpc.html*.

3. John Muir, "The Wild Parks and Forest Reservations of the West," *Atlantic Monthly* 81 (January 1898):15.

A BRIEF ILLUSTRATED HISTORY OF WILDERNESS TIME ~ DOUGLAS W. SCOTT

1. "How near to good…": Brooks Atkinson, ed., "Walking," *Walden and Other Writings of Henry David Thoreau* (New York: The Modern Library, 1950), p. 615; "Wilderness is the raw material…": Aldo Leopold, *A Sand County Almanac and Sketches Here and There* (New York: Oxford University Press, 1949), p. 188; "…the tonic of wildness": Brooks Atkinson, ed., "Spring," *Walden and Other Writings of Henry David Thoreau* (New York: The Modern Library, 1950), p. 283; "life would stagnate…": Ibid.; "We can never have enough…": Ibid.; The classic study of the intellectual evolution of the idea of wilderness, and the contributions of artists and writers, is Roderick Nash, *Wilderness and the American Mind* (New Haven: Yale University Press, 1967). The third edition (1982) is available in paperback.

2. As quoted in Linnie Marsh Wolfe, *Son of the Wilderness: The Life of John Muir* (New York: Knopf, 1945), p. 293.

3. "Wilderness is a necessity.": John Muir, "The Wild Parks and Forest Reservations of the West," *Atlantic Monthly* 81 (January 1898):15; "going to the mountains.": Ibid.; "fountains of life,": Ibid.; "thousands of God's wild blessings": Ibid., p. 24; "climb the mountains…": Linnie Marsh Wolfe, ed., *John of the Mountains: The Unpublished Journals of John Muir* (Boston: Houghton Mifflin, 1938), p. 82; "I never before…": John Muir to L. S. Muir, May 19, 1903, and to the Merriams and the Baileys, January 1, 1904, John Muir Papers, University of the Pacific; "should afford perpetual protection…": As quoted in U.S. Senate, *Establishing a National Wilderness Preservation System for the Permanent Good of the Whole People, and for Other*

Purposes, April 3, 1963, S. Rept. 88–109, p. 20; {Even as Americans…}: Robert Marshall, "The Universe of the Wilderness Is Vanishing," *Nature Magazine* 9, no. 4 (April 1937); {At the turn of the nineteenth century…}: Stephen Fox, *John Muir and the Sierra Club: The American Conservation Movement* (Boston: Little Brown, 1981). A University of Wisconsin Press paperback is available; {Theodore Roosevelt…}: Roderick Nash, *Wilderness and the American Mind* (New Haven: Yale University Press, 1967).

4. As quoted in Curt Meine, *Aldo Leopold: His Life and Work* (Madison: University of Wisconsin Press, 1991), p. 232.

5. "We should keep here and there…": As quoted in Aldo Leopold, "Conserving the Covered Wagon," *Sunset*, March 1925, p. 56; "unimpaired for the enjoyment of future generations": *An Act to Establish a National Park Service, and for Other Purposes*, August 25, 1916, 39 Stat. 535; {In 1872 Congress established…}: Richard West Sellars, *Preserving Nature in the National Parks: A History* (New Haven: Yale University Press, 1997); {Therefore, it wasn't all that surprising…}: The most detailed treatment of the evolution of Forest Service wilderness protection and wilderness policy is James P. Gilligan, "The Development of Policy and Administration of Forest Service Primitive and Wilderness Areas in the Western States" (Ph.D. dissertation, University of Michigan, 1953). A more readily available summary that brings the history up to the late 1980s is Dennis M. Roth, *The Wilderness Movement and the National Forests* (College Station, TX: Intaglio Press, 1988). Roth was the historian of the U.S. Forest Service. One of several biographies of Leopold is Curt Meine, *Aldo Leopold: His Life and Work* (Madison: University of Wisconsin Press, 1991).

6. Robert Marshall, *Alaska Wilderness: Exploring the Central Brooks Range*, George Marshall, ed. (Berkeley: University of California Press, 1956), p. 165.

7. "The harsh environment…": Robert Marshall, "The Problem of the Wilderness," *The Scientific Monthly*, February 1930, p. 143; "it may be a better boast…": Robert Marshall, "A Plea for the Old Wilderness," *The New York Times Magazine*, April 25, 1937, p. 17; {In the 1930s Robert Marshall…}: James M. Glover, *A Wilderness Original: The Life of Bob Marshall* (Seattle: The Mountaineers, 1986).

8. "From the eternity of the past…": Howard Zahniser, Testimony in *Wilderness Preservation System*, Hearings before the House Subcommittee on Public Lands, Committee on Interior and Insular Affairs, (88th Congress, 2d Session) on H.R. 9070, H.R. 9162, S. 4, and Related Bills, April 27–30, and May 1, 1964, p. 1205; "There is no assurance…": Kenneth A. Reid, "Let

Them Alone!," *Outdoor America* 5, no. 1 (November 1939):6; "Let us try to be done...": Howard Zahniser, "How Much Wilderness Can We Afford to Lose?" In *Wildlands in Our Civilization*, David Brower, ed. (San Francisco: Sierra Club, 1964), p. 51; {Into the breach stepped Howard Zahniser...}: Zahniser's role is summarized by Roderick Nash, *Wilderness and the American Mind* (New Haven: Yale University Press, 1967).

9. "We'll hand them a tool...": Rep. Wayne N. Aspinall, May 1954, quoted in "What Is Your Stake in Dinosaur?" In pamphlet published by Trustees for Conservation, San Francisco (1955); {The ensuing fight over the Echo Park Dam...}: The Echo Park fight is splendidly recounted in Mark W. T. Harvey, *A Symbol of Wilderness: Echo Park and the American Conservation Movement* (Albuquerque: University of New Mexico Press, 1994). This is available in paperback from the University of Washington Press. Professor Harvey is now completing a biography of Howard Zahniser.

10. "Wilderness is the original...": Holmes Ralston III, "The Wilderness Idea Reaffirmed," *The Great New Wilderness Debate*, J. Baird Callicott and Michael P. Nelson, eds. (Athens: The University of Georgia Press, 1998), p. 375; All other quotes in this section from *An Act to Establish a National Wilderness Preservation System for the Permanent Good of the Whole People, and for Other Purposes*, September 3, 1964, 78 Stat. 890. See Appendix; {The Wilderness Bill, still a very long shot...}: The legislative campaign to pass the Wilderness Act is well summarized by Roderick Nash, *Wilderness and the American Mind* (New Haven: Yale University Press, 1967), and by political scientist James L. Sundquist, *Politics and Policy: The Eisenhower, Kennedy, and Johnson Years* (Washington, D.C.: Brookings Institution, 1968); {The Act made it the national policy...}: Statistics about the National Wilderness Preservation System and details about each of its more than 640 areas can be found at *www.wilderness.net/NWPS/*.

11. "A great liberating force...": Stewart M. Brandborg, "Executive Director's Report to Council 1966–67," Wilderness Society Papers, Western History Department, Denver Public Library, Box 2. Cited in Dennis M. Roth, *The Wilderness Movement and the National Forests* (College Station, TX: Intaglio Press, 1988), pp. 13–14; {Having Congress vote on each addition...}: The role of Stewart Brandborg and the early years of implementation of the Wilderness Act, especially for the National Forests, are summarized by Dennis M. Roth, *The Wilderness Movement and the National Forests*.

12. Jimmy Carter, "Make This Natural Treasure a National Monument," *The New York Times*, December 29, 2000.

13. "Alaska wilderness areas…": President Jimmy Carter's remarks at signing ceremony, *Alaska National Interest Lands Conservation Act*, December 2, 1980; {The Wilderness Act itself designated no wilderness areas…}: The history of the Alaska lands issue is summarized by Claus M. Naske and Herman E. Slotnick, *Alaska: A History of the 49th State* (Norman, OK: University of Oklahoma Press, 2nd edition, 1987), Chapters 11 and 13.

14. "The de facto wilderness…": David R. Brower, "De Facto Wilderness: What Is Its Place?" In *Wildlands in Our Civilization*, p. 109; "been set aside by God…": Ibid., p. 103; {But the Wilderness Act left the door open…}: The struggles over de facto wilderness on the national forests is recounted by Dennis M. Roth, *The Wilderness Movement and the National Forests*; {Eventually many wilderness area proposals on national forests…}: Frank Wheat, *California Desert Miracle: The Fight for Desert Parks and Wilderness*," (San Diego: Sunbelt Publications, 1999).

15. *An Act to Establish a National Wilderness Preservation System for the Permanent Good of the Whole People, and for Other Purposes*, September 3, 1964, 78 Stat. 890. See Appendix.

American Indians and the Wilderness
~ Vine Deloria, Jr.

1. Roderick Nash, *Wilderness and the American Mind*, Rev. ed. (New Haven: Yale University, 1973), p. 26.

2. Luther Standing Bear, *Land of the Spotted Eagle* (New York: Houghton Mifflin, 1932), p. xix.

3. Harvey Cox, *The Secular City* (New York: Macmillan, 1965), p. 20.

4. Luther Standing Bear, *Land of the Spotted Eagle*, p. 196.

5. Ibid., p. 248.

6. Ibid.

7. Alexis de Tocqueville, *Democracy in America* (New York: Doubleday Anchor, 1969), p. 322.

8. Ibid., p. 323.

9. Ibid.

10. Luther Standing Bear, *Land of the Spotted Eagle*, p. 166.

11. Ibid., p. 194.

THE FIRST CONSERVATIONISTS
~ CHRIS MADSON

1. Jean-Marie Chauvet, Eliette Brunel Deschamps, and Christian Hillaire, *Dawn of Art: The Chauvet Cave, The Oldest Known Paintings in the World* (New York: Harry N. Abrams, Inc., 1996), p. 122.

2. Elaine Anderson, "Who's Who in the Pleistocene: A Mammalian Bestiary," Paul S. Martin and Richard Klein, eds., *Quaternary Extinctions: A Prehistoric Revolution* (Tucson: University of Arizona Press, 1989), pp. 40–89.

3. Christopher Stringer and Clive Gamble, *In Search of the Neanderthals* (New York: Thames and Hudson, Inc., 1993), p. 181, 195, 201.

4. Jean Clottes and David Lewis-Williams, *The Shamans of Prehistory: Trance and Magic in the Painted Caves* (New York: Harry N. Abrams, Inc. 1996), pp. 41–49.

5. Ibid., pp. 61–114.

6. Ibid., pp. 66–72, and Mario Ruspoli, *The Cave of Lascaux: The Final Photographs* (New York: Harry N. Abrams, Inc., 1987), p. 149.

7. Mario Ruspoli, *The Cave of Lascaux: The Final Photographs*, p. 89.

8. Goren Burenhult, ed., *People of the Stone Age: Hunter-gatherers and Early Farmers*. Vol. 2 of *The Illustrated History of Humankind, American Museum of Natural History*, (San Francisco: HarperCollins Publishers, 1993), pp. 22–32.

9. Paul Shepard, *Coming Home to the Pleistocene* (Washington, D.C.: Island Press, 1998), pp. 81–108. See also such Biblical references as Leviticus 26: 6: "And I will give peace in the land and ye shall lie down and none shall make you afraid: and I will rid evil beasts out of the land…" and Deuteronomy 1:19: "And when we departed from Horeb we went through all that great and terrible wilderness…" See also the *Epic of Gilgamesh* in which Enkidu comes on the scene as an all-powerful giant that socializes with wild beasts and terrifies the local populace. Not even Gilgamesh can deal with him until he is weakened by sexual liaison with a prostitute. See also the details of art from ancient civilizations in Greece and the Fertile Crescent—start with some of the tombstone art out of Grave Circle B at

Mycenae, circa 1600 B.C. These show lions attacking cattle and hunters. These are all shadows of ancient traditions among Middle Eastern agrarian peoples.

10. Ibid.

11. Paul Shepard, *Coming Home to the Pleistocene*, pp. 81–108.

12. Goren Burenhult, ed., *People of the Stone Age: Hunter-gatherers and Early Farmers*, p. 29.

13. Homer, *The Odyssey*, trans. Robert Fitzgerald (Garden City, NY: Doubleday & Company, Inc., 1963), Lines 442–461, p. 367.

14. John Kinloch Anderson, *Hunting in the Ancient World* (Berkeley and Los Angeles: University of California Press, 1985), p. 9.

15. See Edward I's confirmation of the Charter of the Forest, March 28, 1299, an addendum to the *Magna Carta*.

16. Aldo Leopold, *Game Management* (New York and London: Charles Scribner's Sons, 1933), p. 7.

17. William Bradford, *History of the Plimouth Plantation, http://members.aol.com/ calebj/bradford_journal9.html*.

18. Jack Dempsey, ed., *The New English Canaan*, by Thomas Morton, 1637 (Scituate, MA: Digital Scanning, Inc., 1999), pp. 53–54.

19. Robert Elman, *First in the Field* (New York: Mason/Charter, 1977), pp. 9–25.

20. Maria R. Audubon, *Audubon and His Journals: Volume II* (Mineola, NY: Dover Publication, Inc., 1986 (reprinted from the original, published in 1897 by Scribner, New York, NY, 1986)), entry for Saturday, August 5, 1843, p. 131.

21. John F. Reiger, *American Sportsmen and the Origins of Conservation* (Norman: University of Oklahoma Press, 1975), p. 48.

22. Frank Graham, Jr., *The Adirondack Park: A Political History* (New York: Alfred A. Knopf, 1978), p. 59.

23. Ibid., pp. 59–60, and Craig W. Allin, *The Politics of Wilderness Preservation* (Westport, CT: Greenwood Press, 1982), p. 25.

24. John F. Reiger, *American Sportsmen and the Origins of Conservation*.

25. Ibid.

26. Frank Graham, Jr., *The Adirondack Park: A Political History*, pp. 85–106.

27. Ibid., p. 120, and Roderick Nash, *Wilderness and the American Mind*, (New Haven: Yale University Press, 1967), p. 133, and John F. Reiger, *American Sportsmen and the Origins of Conservation*.

28. Edmund Morris, *The Rise of Theodore Roosevelt* (New York: Coward, McCann, & Geoghegan, Inc., 1979).

29. James B. Trefethen, *An American Crusade for Wildlife* (Alexandria, VA: Boone and Crockett Club, 1975), pp. 117–128.

30. Roderick Nash, *Wilderness and the American Mind* (New Haven: Yale University Press, 1967), p. 183, and John F. Reiger, *American Sportsmen and the Origins of Conservation*, p. 78, and Curt Meine, *Aldo Leopold: His Life and Work* (Madison: University of Wisconsin Press, 1988), p. 78.

31. Aldo Leopold, *A Sand County Almanac* (London: Oxford University Press, 1975), p. 130.

32. Ibid., pp. 148–149.

33. Donald N. Baldwin, *The Quiet Revolution: Grass Roots of Today's Wilderness Preservation Movement* (Boulder, CO: Pruett Publishing Company, 1972), p. 34.

34. Ibid., p. 100–106.

35. Curt Meine, *Aldo Leopold: His Life and Work* (Madison: University of Wisconsin Press, 1988), p. 224.

36. Aldo Leopold, *Aldo Leopold's Wilderness: Selected Early Writings by the Author of* A Sand County Almanac, David E. Brown and Neil B. Carmony, eds., (Harrisburg, PA: Stackpole Books, 1990), p. 159.

37. Curt Meine, *Aldo Leopold: His Life and Work*, p. 343.

38. Aldo Leopold, *Living Wilderness*, September 1935.

39. Aldo Leopold, *A Sand County Almanac and Sketches Here and There* (New York: Oxford University Press, 1949), p. 200.

40. The Wilderness Society website, *http:wcs.org/home/wild/northamerica/anwr/7895* and *http://www.r7.fws.gov/mwr/arctic/descrip.html*. See also *Who Was Who in America: Volume IV, 1961–1968* (Chicago: Marguis—Who's Who, 1968), p. 689.

41. *Arizona Republic*, June 4, 1967, *http://dizzy.library.arizona.edu/branches/spc/sludall/articleretrievals/inside.html*.

42. The Wilderness Society website, *http://pewwildernesscenter.org/alaska.htm*.

43. Dennis M. Roth, *The Wilderness Movement and the National Forests 1964–1980* (U.S. Forest Service Publication FS-391, 1984), cited on p. 21, Orion The Hunter's Institute, *1999 Annual Report, Orion The Hunter's Institute.*

44. Finis Mitchell, *Wind River Trails: A Hiking and Fishing Guide to the Many Trails and Lakes of the Wind River Range in Wyoming* (Salt Lake City: Wasatch Publishers, Inc., 1975), p. 142.

45. Tom Reed, personal communication, March 26, 2001.

46. Aldo Leopold, *A Sand County Almanac and Sketches from Here and There* (New York: Oxford University Press, 1949), p. 225.

47. Aldo Leopold, *Round River, from the Journals of Aldo Leopold* (New York: Oxford University Press, 1953), p. 155.

48. Ron Marlenee, March 4, 1999. Testimony of Safari Club International before the Subcommittee on Forest and Forest Health, Committee on Resources, U.S. House of Representatives. *http://resourcescommittee.house.gov/resources/106cong/forests/99mar04/marlenee.htm*.

49. American Hunter Online. Downloaded March 22, 2001 from *http://american-hunter.org*.

50. Craig Thomas, June 28, 2000, "Senator Stakes out Public Access to Challenge Forest Plan," *http://thomas.senate.gov.html/pr289.html*.

51. Lance Morrow, September 11, 2000, "The Perfect Firestorm," *Time Canada*, 156 (11), *www.canoe.ca/TimeCanada0009/11_time27.html*.

52. Eileen King, nd., "Survey shows broad support among hunters and anglers for retaining roadless areas and access in national forests." *www.trca.org/ news/default.asp?news_id=126.*

53. Chris Potholm, March 24, 2001. Interview.

54. Ted Kerasote, "The Theodore Roosevelt Conservation Alliance," *Sports Afield*, August 2001.

55. C. H. D. Clarke, 1975, "Venator—The Hunter." Unpublished manuscript of a speech. No information on the date or location of the presentation.

"GIFTS OF NATURE" IN AN ECONOMIC WORLD
~ THOMAS MICHAEL POWER

1. Southwich Associates, 2000, "Historical Economic Performance of Oregon and Western Counties Associated with Roadless and Wilderness Areas," p. 7.

2. Ibid. The correlation coefficients for the most rural counties (no city greater than 2,500) were 0.33 and highly significant. The correlation was also significant for all western counties as well as all western non-metropolitan counties. The state included were Montana, Wyoming, Colorado, New Mexico, Idaho, Utah, Arizona, Nevada, California, Oregon, and Washington.

3. Paul Lorah, "Population Growth, Economic Security, and Cultural Change in Wilderness Counties." In *Wilderness Science in a Time of Change Conference*, by David N. Cole, et al., eds. (Fort Collins, CO: U.S. Department of Agriculture, Forest Service, Rocky Mountain Research Station, 2000), RMRS-P-15-CD.

4. David H. Jackson and Kenneth Wall, "Mapping and Modeling Real Estate Development in Rural Western Montana" (Missoula, MT: Bolle Center for People and Forests, School of Forestry, University of Montana, 1995), Discussion Paper No. 2.

5. Spencer Phillips, "Windfalls for Wilderness: Land Protection and Land Value in the Green Mountains." In *Wilderness Science in a Time of Change Conference*, by David N. Cole et al., eds., and Spencer Phillips, *Windfalls for Wilderness: Land Protection and Land Value in the Green Mountains* (Craftbury Common, VT: Ecology and Economics Research Group, Wilderness Society, 1999).

6. David Lewis and Andrew J. Plantinga, "Public Conservation Lands and Economic Growth in the Northern Forest Region" (Orono ME: Department of Resource Economics and Policy, University of Maine, November 17, 2000), p. 12–13.

7. Kevin T. Duffy-Deno, "The Effect of Federal Wilderness on County Growth in the Intermountain Western United States." In *Journal of Regional Science,* 1998, 38(1):109–136.

8. Gundars Rudzitis and Rebecca Johnson, "The Impact of Wilderness and Other Wildlands on Local Economies and Regional Development Trends." In *Wilderness Science in a Time of Change Conference,* by David N. Cole et al., eds., and Gundars Rudzitis, *Wilderness and the Changing American West* (New York: John Wiley and Sons, 1996), Figure 7.1 and pp. 112–116.

9. John L. Crompton et al., "An empirical study of the role of recreation, parks and open space in companies' (re)location decisions." In *Journal of Park and Recreation Administration,* 1997, 15(1):37–58; J. Johnson and R. Rasker, "The Role of Amenities in Business Attraction and Retention." In *Montana Policy Review* 1993, 3(2):11–19; Ray Rasker, "A New Look at Old Vistas: The Economic Role of Environmental Quality in Western Public Lands." In *University of Colorado Law Review,* 1994, 65(2):369–397; Ashish Arora, Richard Florida, Gary J. Gates and Mark Kamlet, "Human Capital, Quality of Place, and Location" (Pittsburgh: H. John Heinz School of Public Policy at Carnegie Mellon University, 2000); Paul D. Gottlieb, "Residential Amenities, Firm Location and Economic Development." In *Urban Studies,* 1995, 32(9):1413.

10. Richard Florida, *Competing in the Age of Talent: Environment, Amenities, and the New Economy* (H. John Heinz III School of Public Policy and Management, Carnegie Mellon University. A report prepared for the R. K. Mellon Foundation, Heinz Endowments, and Sustainable Pittsburgh, 2000, p. 5.

11. If income data is used instead of employment data, the upward trend caused by inflation has to be removed first by deflating the dollar figures by an appropriate measure of inflation, *e.g.* the Consumer Price Index or the GNP deflator for consumer expenditures.

12. This is an important point upon which economists' and popular opinion often significantly diverge. Workers who are laid off almost immediately begin seeking alternative employment. The number who become re-employed steadily increases with each passing month. Within a year over

90 percent will be re-employed and the unemployment rate among that group will be about the same as for the workforce as a whole (Bureau of Labor Statistics, 1988). Neither the nation nor any region has faced steadily increasing unemployment rates over significant periods of time. Laid off workers, in general, shift to alternative employment. Capital moves even more quickly to alternative uses. Financial balances, for instance, do not even sit idle over night. Money markets allow them to be lent out for fractions of a day. Land, because it is not mobile in the way labor and capital are, may take a longer period before it is re-employed, but falling rental or sales prices will work toward attracting alternative uses for land too.

13. U.S. Dept. Comm., BEA, REIS.

14. Dave Larsen and Phil Aust, "Has the Sun Really Set on Washington State's Forest Products' Industry?" (Washington Mill Survey, 1986–1996, Washington State Department of Natural Resources, April 26, 2000. A paper prepared for the 34th Annual Pacific Northwest Regional Economic Conference).

15. E. Niemi and E. Whitelaw, "Why the Sky Did Not Fall in the Pacific Northwest," (Eugene, Oregon: EcoNorthwest, 1998), p. 61, employment growth from USDOC, BEA, REIS CD-ROM.

16. For an example of the application of the type of employment analysis discussed below see Thomas Michael Power, "To Be or Not To Be: The Economics of Natural Gas Development Along the Rocky Mountain Front." In *Western Wildlands*, fall 1987, pp. 20–25.

THE POLITICS OF PROTECTING WILD PLACES
~ MIKE MATZ

1. See appendix, "Wilderness Act of 1964," Public Law 88–577 (16 U.S.C. 1131-1136), Section 2(c).

2. 66 Federal Register 3244, Notice of Final Rule, 36, CFR Part 294, January 12, 2001. See also "Forest Service Roadless Area Conservation," Final Environmental Impact Statement, November 2000. The Forest Service received more than 1.6 million comments on the proposed rule, the most comments for any rulemaking in history.

3. For a list of presidential proclamations by Clinton establishing national monuments, see *http://www.access.gpo.gove/nara/nara003.html*.

4. U.S. Census Bureau, 2000 census figures, Table 5, "Metropolitan Areas

Ranked by Percentage Population Change: 1990–2000," at *http://www.census.gov/population/cen2000/phc-t3/tab05.pdf*, April 2, 2001. Las Vegas grew by 83.3 percent over the last decade, outpacing the second-fastest growing city, Naples, FL, at 65.3 percent by almost 20 percent.

5. See Natural Resources Defense Council website report on Gale Nortorn, *www.nrdc.org/legislation/norton/nortoninx.asp.*

6. See *ABC News* website profiles on Ann Veneman, *http://abcnews.go.com/sections/politics/DailyNews/profile_veneman.html.*

7. Editorial, *Atlanta Constitution*, August 10, 2001.

8. State of Idaho vs. U.S. Forest Service, U.S. District Court, District of Idaho, Case No. CV01-11-N-EJL, Order by Judge Edward Lodge, May 10, 2001.

9. Dan Balz, "Despite Wins, Bush Faces Battles Ahead," *Washington Post*, August 3, 2001, p. A1.

10. "CNN/Time Poll" conducted March 21–22, 2001 by Yankelovich Partners and Harris, 1,025 adults surveyed, margin of error +/- 3.1 percent.

11. "TNS Intersearch" conducted April 19–22, 2001. Surveyed 1,350 adults, margin of error ± 2.5 percent.

12. "ABC News/Washington Post" conducted July 26–30 and released August 1, 2001. Surveyed 1,352 adults, margin of error ± 2.5 percent.

13. Ibid.

14. Dan Balz, "Despite Wins, Bush Faces Battles Ahead."

15. Mark Z. Barabak, "Bush Criticized as Fear for Environment Grows," *Los Angeles Times*, April 30, 2001.

16. Editorial, *Idaho Statesman*, July 31, 2001.

17. Mark Z. Barabak, "Bush Criticized as Fear for Environment Grows."

18. National Survey on Recreation and the Environment, 2000 (NSRE 2000), U.S.D.A. Forest Service and N.O.A.A., Summary Report #2, WILD279, p. 3.

CHRISTIANITY AND WILD PLACES
~ STEVEN BOUMA-PREDIGER

1. Roderick Nash, *The Rights of Nature* (Madison: University of Wisconsin, 1989), p. 90.

2. Lynn White, Jr., "The Historical Roots of Our Ecologic Crisis," *Science* 155 (March 10, 1967), pp. 1203–1207. Also found in Ian Barbour, ed., *Western Man and Environmental Ethics* (Reading, MA: Addison-Wesley, 1973).

3. Lynn White, Jr., "The Historical Roots of Our Ecologic Crisis," in Ian Barbour, ed., *Western Man and Environmental Ethics*, (Reading, MA: Addison-Wesley, 1973), p. 27.

4. Ibid., p. 25.

5. Wendell Berry, *What Are People For?* (New York: North Point, 1990), p. 98.

6. See Ron Wolf, "God, James Watt, and the Public Lands," *Audubon* 83, no. 3 (May 1981), and Robert Booth Fowler, *The Greening of Protestant Thought* (Chapel Hill: University of North Carolina, 1995), p. 47.

7. Roderick Nash, *The Rights of Nature*, pp. 91–92. For more on the "end times" literature, see Fowler, *The Greening of Protestant Thought*, Chapter 3.3.

8. For discussion of this text, see Richard Bauckham, *Word Biblical Commentary 50: Jude and 2 Peter* (Waco: Word, 1983), pp. 303–322.

9. Aldo Leopold, *Sand County Almanac* (New York: Ballantine, 1970), p. 203.

10. Jürgen Moltmann, *God in Creation* (San Francisco: Harper and Row, 1985), p. 139.

11. James Nash, *Loving Nature* (Nashville: Abingdon, 1991), pp. 74–75. The historical work of Carolyn Merchant, *The Death of Nature* (San Francisco: Harper and Row, 1980), and Clarence Glacken, *Traces on the Rhodian Shore* (Berkeley: University of California, 1967), also repudiates any theory of single causation with respect to ecological degradation. For an excellent overview of world history from an environmental perspective, see Clive Ponting, *A Green History of the World* (New York: Penguin, 1991).

12. James Nash, *Loving Nature*, p. 72.

13. Ibid., pp. 72, 74.

14. Wendell Berry, *Sex, Economy, Freedom, and Community* (New York: Pantheon, 1993), p. 94.

15. Ibid., pp. 98, 101.

MOVERS AND SHAKERS IN THE
CHRISTIAN ENVIRONMENTAL MOVEMENT
~ STEVEN BOUMA-PREDIGER

1. Joseph Sittler, "Ecological Commitment as Theological Responsibility," *Zygon* 5 (June 1970):173. For more on Sittler, see *Evocations of Grace: The Writings of Joseph Sittler on Ecology, Theology, and Ethics*, Steven Bouma-Prediger and Peter Bakken, eds. (Grand Rapids, MI: Eerdmans, 2000).

2. Joseph Sittler, *Gravity and Grace* (Augsburg, 1986), p. 15.

WORKING WITH FAITH-BASED ORGANIZATIONS
~ SUELLEN LOWRY

1. Robert D. Putnam, *Bowling Alone* (New York: Simon and Schuster, 2000) p. 70.

2. When talking with individuals in the religious community about communicating with their members of Congress, I often suggest they consider how they would respond if, while conversing with someone in their own living room, they were asked why they care about the environment from a faith-based standpoint. This type of preparation encourages people to talk about what they feel and know best.

3. The importance of relationships to people's commitment and learning can be found in scholarship from many fields, including sociology, psychology, and education. For example, in *Bowling Alone*, Harvard professor Robert D. Putnam emphasizes the importance of "social capital," which is largely produced by engaging people in a face-to-face context. He notes that "[s]ocial networks are the quintessential resource of movement organizers." Putnam 152. In his well-known list of basic needs, Abraham Maslow noted the need to belong (see, *e.g.*, Abraham H. Maslow, *Toward a Psychology of Being* (New York: Von Nostrand Reinhold Company, 1968)). Lev Vygotsky, the Russian developmental theorist and researcher, maintained, in part, that people are most apt to learn when engaged in relational dialogue (see, *e.g.*, Thomas G. Gunning, *Creating Literacy Instruction for All Children* (Needham Heights: Allyn & Bacon, 2000) pp. 3–4).

THE WILD AND ITS NEW ENEMIES
~ JACK TURNER

1. See, for example, the News Release for *State of the World 2001*, published by the Worldwatch Institute, at *www.worldwatch.org*.

2. I confine myself in this essay to the first technology, "gene splicing." Cloning of endangered species is still in its infancy, although there is a growing number of "frozen zoos" holding genes from endangered species to be used by cloning technology. "Digital matter" is one of the names for the product of nanotechnology, by far the most marvelous and dangerous technology yet created. Nanotechnology seeks to create programmable machines at the molecular scale, sometimes called molecular robots or nanobots. Such machines could replicate and self-assemble into complex systems to carry out specific tasks—mining for ore, devouring toxins in a stream, or "building" a tree. There are a number of nanoecology websites that discuss the design and reconstruction of "the ecology" to enable humans to survive radical climate change. K. Eric Drexler discussed the subject in his classic text, *Engines of Creation*, and there is on-going coverage of its development at Drexler's Foresight Institute website: *www. foresite.org*. The Institute advocates cloning of endangered and "extinct" species and the use of nanotechnology to "reconstruct" habitats for them. The very existence of this ideology undermines the defense of biodiversity and wilderness preservation by promoting replacement rather than preservation.

3. The National Nanotechnology Initiative received $422 million in funding in 2001 for research in nanotechnology. For comparison, the U.S. Fish and Wildlife Service's funding for the Endangered Species Act Programs is $121 million for 2001.

4. Phillip S. Anton, Richard Silberglitt, and James Schneider, *The Global Technology Revolution: Bio/Nano/Materials Trends and Their Synergies with Information Technology by 2015,* p. 6. The report is available at *www.rand. org/publications/MR/MR1307/*.

5. *www.oecd.org/ehs/summary.htm*. The function of the field trials varies with the organism being tested, but it usually explores how the transgene expresses a trait or traits in the new organism and whether this expression is useful—to science, medicine, or commerce.

6. *www.oecd.org/ehs/summary.htm*.

7. At *www.nytimes.com* you can watch an excellent video explanation of gene

transfer technology. Go to *www.nytimes.com/pages/health/genetics* and select "gene splicing primer" under "interactive images."

8. Washington, D.C., Associated Press, March 8, 2001.

9. *www.darpa.mil/focus/Bio_Interfaces1.htm.*

10. In 1998 and 1999, scientists at the Australian National University in Canberra accidentally created a mouse virus (a version of mousepox) with gene splicing technology that crippled the immune system of their experimental population of mice, leaving 90 percent of the mice dead and the rest permanently disabled. They published their results, which explained how they did it, in *The Journal of Virology*. See "Australians Create Deadly Mouse Virus," by William J. Broad, *The New York Times*, January 23, 2001. The Rand Corporation report mentioned above notes that unlike other weapon systems a genetically engineered pathogen could be produced by a small business or an individual trained in microbiology.

11. Keekok Lee, *The Natural and the Artefactual: The Implications of Deep Science and Deep Technology for Environmental Philosophy* (New York: Lexington Books, 1999). This is the best treatment of a new subject, a true pioneering work in philosophy; unfortunately, the writing is poor.

12. Robert P. Lanza, Betsy L. Dresser, and Philip Damiani, "Cloning Noah's Ark," *Scientific American*, November 2000.

13. "Exotic Species Invading Yellowstone," *Yellowstone Discovery* 16, no. 1 (spring 2001):4.

14. Margot Higgens, "Aquaculture business booming, but at what cost?" Environmental News Network, February 23, 2001.

15. The Associated Press, "Most North American Fish Species are Endangered," November 3, 2000.

16. "Aquaculture Wed With Biotechnology Seen As Happy Marriage," Earth Vision Environmental News Network, October 2, 2000, *www.earthvision.net.*

17. Dr. Jan van Aken, "Genetically Engineered Fish: Swimming Against the Tide," Greenpeace Media Center, January 2000, *www.greenpeace.org/media/publications/genetic.*

18. Greenpeace, Press Release, March 27, 2001.

19. "Wild Fish Face More Threats," *Anchorage Daily News*, August 29, 2000.

20. Rajiv Sekhri, "Greenpeace seeks curbs on genetically modified fish," Environmental News Network, March 28, 2001.

21. The resolution was adopted in June 2000. It is available on the Greenpeace website, *www.greenpeace.org/~geneng/highlights/gmo/asih.pdf.*

22. Greenpeace, Q&A, *www.greenpeace.org.*

23. "Wild Fish Face More Threats."

24. Beth Daley, "Genetic Pollution: Escaped Farm Salmon Raise Alarm in Maine," *Boston Globe*, February 23, 2001.

25. W. M. Muir and R. D. Howard, "Possible ecological risks of transgenic organism release when transgenes affect mating success: sexual selection and the Trojan gene hypothesis," *Proceedings of the National Academy of Sciences* 96 (1999):13853-13856.

26. See "Reproductive manipulation of fishes; ecologically safe assessment of introduction," by W. L. Shelton, US-ARS, Biotechnology Risk Assessment Research Grants, Program Abstract of Funded Research 1996.

27. Dr. Jan van Aken, "Genetically Engineered Fish."

28. Genetix Alert News Release, March 23, 2001, *http://riseup.net/pipermail/v-mgj/2001-march/000124.html.*

29. Rick Weiss, "Biotech Research Branches Out: Gene-Altered Trees Raise Thickets of Promise, Concern," *The Washington Post*, August 3, 2000.

30. Sherman Paul, *The Shores of America: Thoreau's Inward Exploration* (Urbana: University of Illinois Press, 1958), pp. 412–417; and Jay Hansford C. Vest, "Will of the Land," *Environmental Review*, Winter 1985.

31. Eric Partridge, *Origins: A Short Etymological Dictionary of Modern English* (New York: Macmillan Company, 1958), p. 806.

32. Rick Weiss, "Biotech Research Branches Out."

33. Ibid.

34. Jon R. Luoma, *The Hidden Forest: The Biography of an Ecosystem* (New York: Henry Holt, 1999), p. 38.

35. James C. Scott, *Seeing Like a State: How Certain Schemes to Improve the Human Condition Have Failed* (New Haven: Yale University Press, 1998), p. 13.

36. Ibid., p. 20.

37. Jon R. Luoma, *The Hidden Forest*, p. ix.

38. Rick Weiss, "Biotech Research Branches Out."

39. See Corner House Briefing No. 15, "The Dyson Effect: Carbon 'Offset' Forestry and the Privatization of the Atmosphere," *http://cornerhouse.icaap. org/briefings/15.html.*

40. Ibid.

41. Ibid.

42. Rick Weiss, "Biotech Research Branches Out."

43. Ibid.

44. Elizabeth Becker, "New Worries of Planting Altered Corn," *The New York Times,* March 2, 2001.

45. Corner House Briefing No. 21, "Genetic Dialectic: the Biological Politics of Genetically Modified Trees." The best discussion so far of transgenic forestry.

46. Ibid.

47. Ibid.

48. David Lukas, "Of Aerial Plankton and the Aeolian Zone," *Orion,* 18, No. 2, Spring 1999.

49. Corner House Briefing No. 21.

50. Fen Montaigne, "A River Dammed," *National Geographic,* April 2001.

51. Jon R. Luoma, *The Hidden Forest*, p. 9.

52. Corner House Briefing No. 21.

SHOULD WILDERNESS BE MANAGED?
~ MICHAEL E. SOULÉ

1. Humanitarian (social justice) arguments for the eternal and essential symbiosis of humanity and wilderness have been made, of course, although these contradict the spirit of the Wilderness Act of 1964. In any case, such discourse is more relevant to "The Great New Wilderness Debate" (J. B. Callicott and M. P. Nelson, *The Great New Wilderness Debate* (Athens and London: University of Georgia Press, 1998)) than to the issue at hand, which is the role of management within designated wilderness.

2. Jack Turner, "Jack Turner: *The Abstract Wild.*" An interview in *Wild Duck Review* 2, no. 6, pp. 9–13; Jack Turner, *The Abstract Wild* (Tucson: The University of Arizona Press, 1996). By the way, the science is tested and is standing up well (see Note 6).

3. The degradation and diminution of terrestrial, aquatic, and marine ecosystems, while tragic, is the stuff of future myths. Whereas the first fall—the banishment from Eden—represents the ego's triumph over the soul and the origin of Nature-humanity dualism—the mythical and social implications of the current fall—the physical destruction of Nature—has yet to crystallize. Some people experience the conquest of Nature as a sign of progress, civilization's victory over darkness, over the Goddess, and the fulfillment of Yahweh's desire to subdue fecund wildness.

4. See references in M. E. Soulé, 2000/2001. Does sustainable development help nature? *Wild Earth,* winter 2000/2001, pp. 56–64.

5. Other perspectives are not shown here because they are not central to the debate on wilderness management among North American conservationists. Among the views are (1) the Wise Use movement's perversion of Gifford Pinchot's ethic of utilitarianism and (2) the social justice movement's emphasis on the welfare of those who lack much power. Both of these movements have adherents who are motivated by good will toward nature and humanity though both tend to be dominated by extremists or those who believe that a wildland is wasted unless it benefits people economically.

6. P. Landres, M. W. Brunson, and L. Merigliano, "Naturalness and Wildness," *Wild Earth,* winter 2000/2001, pp. 77–82.

7. M. E. Soulé and R. K. Noss, "Rewilding and biodiversity as complementary

tools for continental conservation," *Wild Earth*, fall 1998, pp. 18–28; M. E. Soulé and J. Terborgh, eds., *Continental Conservation: Scientific Foundations of Regional Reserve Networks* (Washington, D.C. and Covelo, CA: Island Press, 1999).

8. P. Landres, M. W. Brunson, and L. Merigliano, "Naturalness and Wildness."

9. For some social critics, however, the creation of enemies is a lesser sin than quietism in the face of an evil system requiring a social revolution to change (*e.g.*, Jack Turner references in Note 2.)

10. The endangerment of the condor was solely anthropogenic—caused by land use changes, management abuses, shooting of condors, and the use of lead bullets in small game and big game hunting, the fragments of which were ultimately ingested by condors; lead is often lethal to condors because it arrests peristalsis. Incidentally, it is a myth that the California condor is an evolutionary "dead end" or a senescent species that was already on the way out; its dire situation is entirely anthropogenic, although the human actions contributing to its status probably began with the early Holocene slaughter of large mammals in North America.

11. Jack Turner, "Jack Turner: *The Abstract Wild.*"

12. Ideally, the wolf would have made it back on its own, but the complex politics of Wyoming and Montana, and the hostility of ranchers and others in the region, justified using the shield of the Endangered Species Act in a less than ideal manner to implement the introduction.

13. W. R. Ripple and E. J. Larsen, "Historic aspen recruitment, elk, and wolves in northern Yellowstone National Park," USA. *Biological Conservation* 95 (2000):361–370.

14. P. Singer, *Animal Liberation* (New York: Avon Books, 1990); Jack Turner, Forum: Grizzly bear number 209. *Wild Duck Review* 1997, 3(1):11–13.

15. G. Nickas, "Wilderness Fire," *Wilderness Watcher* 10 (1998):3–5; G. Nickas and G. Macfarlane, "Ecological Restoration in Wilderness." *Wild Earth*, in press.

16. For a recent review of wilderness restoration see D. Simberloff, D. Doak, M. Groom, S. Trombulak, A. Dobson, S. Gatewood, M. E. Soulé, M. Gilpin, C. Martinez Del Rio, and L. Mills, "Regional and Continental Restoration." In M. E. Soulé and J. Terborgh, eds., *Continental Conservation: Scientific*

Foundations of Regional Reserve Networks, pp. 65–98.

17. R. F. Noss, E. Dinerstein, B. Gilbert, M. E. Gilpin, B. Miller, J. Terborgh, and S. Trombulak, "Core Areas: Where Nature Reigns." In M. E. Soulé, J. Terborgh, eds., *Continental Conservation: Scientific Foundations of Regional Reserve Networks,* pp. 99–128.

18. See Note 4.

19. M. E. Soulé, "Thresholds for survival: criteria for maintenance of fitness and evolutionary potential." In M. E. Soulé and B. M. Wilcox, eds., *Conservation Biology: An Evolutionary-Ecological Perspective* (Sunderland, MA: Sinauer Associates, 1980), pp. 151–170. For comparison, the total area of Yellowstone and Grand Teton National Parks in Wyoming is less than 10,000 square kilometers.

20. Ibid.

21. R. F. Noss and A. Y. Cooperrider, *Saving Nature's Legacy* (Washington, D.C. and Covelo, California: Island Press, 1994), p. 172.

22. P. B. Landres, "Temporal scale perspectives in managing biological diversity," *Transactions of the North American Wildlife and Natural Resources Conference* 57 (1992):292-307.

23. Along the lines proposed by The Wildlands Project. D. Foreman, "The Wildlands Project and the Rewilding of North America." In *Denver University Law Review* 76 (1999):535–553; Private lands will also be essential in many places, and incentives such as conservation easements are encouraging private land owners to become allies of conservation without compromising private property rights.

24. G. H. Boettner, J. S. Elkinton, and C. J. Boettner, "Effects of a biological control introduction on three nontarget native species of Saturniid moths," *Conservation Biology* 14 (2000):1798-1806.

25. See Note 6.

26. See Note 6.

27. C. A. Sydoriak, C. D. Allen, and B. F. Jacobs, "Would landscape restoration make the Bandelier Wilderness more or less of a wilderness?" *Wild Earth,* winter 2000/2001, pp. 83–90.

28. Ibid.

29. Ibid.

30. See Note 2.

MARKETING THE IMAGE OF THE WILD
~ HAL HERRING

1. Joe Kolman, *Missoulian*, December 13, 1999.

2. Ted Williams, "The Elk Ranching Boom," *Audubon*, June 1992.

3. For a full discussion of CWD see, *Transmissible Spongiform Encephalopathies*, Council for Agricultural Science and Technology Task Force Report, No. 136 (October 2000), available as a PDF file at */www.cast-science.org/pdf/tse.pdf*.

4. Sandra Blakeslee, "Biologists Say Hunters Should Beware of Brain Disease," *The New York Times*, October 21, 2000.

5. Ted Kerasote, "The Great Game Ranching Debate," *Sports Afield*, April 2001, p. 93.

THE ONCE AND FUTURE GRIZZLY
~ TODD WILKINSON

1. William Henry Wright, *The Grizzly Bear* (New York: Scribner, 1909), p. 68.

2. Interview with Hank Fischer, January 29, 2001 and on February 15, 2001. Note: Unless otherwise identified with endnotes, all other quotations are based on interviews I conducted with the subjects.

3. David Quammen, *The Song of the Dodo* (New York: Scribner, 1996). Other excellent references on island biogeography and the faster rate of species extinction on islands include: John Terborgh, "Preservation of Natural Diversity: the Problem of Extinction Prone Species," *BioScience* 24, 1974; M. E. Soulé, "What Do We Really Know About Extinction" in *Genetics and Conservation*, Christine Schonewald-Cox, Steven M. Chambers, Bruce MacBryde, and W. Lawrence Thomas, eds. (Menlo Park, CA: Benjamin/Cummings, 1983).

4. Mark Shaffer, *Keeping the Grizzly Bear in the American West: A Strategy for Real Recovery* (Washington, D.C.: The Wilderness Society, 1992). Also notes by Mark Shaffer and Fred Samson, "Population Size and Extinction: A Note on Determining Critical Population Sizes," *American Naturalist* 125,

no. 1, 1985.

5. Hank Fischer, "Bears and the Bitterroot," *Defenders Magazine*, winter 1996–1997, pp. 16–27. See also an earlier story written by Fischer in *Defenders Magazine*, winter 1993–1994, and Hank Fischer and Michael Roy, "New Approaches to Citizen Participation in Endangered Species Management: Recovery on the Bitterroot Ecosystem," *Ursus* 10, 1998.

6. Interview conducted with Bruce Babbitt in Yellowstone in 1995.

7. Tom Kenworthy, "Unlikely Alliance Funds Common Ground for Grizzlies," *The Washington Post*, October 29, 1995.

8. Record of Decision and Final Rule for Grizzly Introduction in the Bitterroot, the Fish and Wildlife Service, which includes a list of alternatives addressed in the Environmental Impact Statement. Written by Dr. Christopher Servheen, released to the public in November 2000. See also a letter written by Montana Governor Marc Racicot on July 18, 1995 to John Weaver who was heading up the Environmental Impact Statement. (Letter available from Defenders of Wildlife Northern Rockies Office, 406/549-0761.) The letter is significant because it demonstrates that an influential westerner and present advisor to President George W. Bush believes the citizen's plan provides a "superb opportunity for local citizen management of grizzly reintroduction." He adds, alluding to the federal conservation mandates handed down from Washington: "The Coalition's approach represents the kind of Endangered Species Act flexibility and the local partnership concepts that the Secretary of the Interior [Bruce Babbitt] has been advocating."

9. Aldo Leopold, "Threatened Species," 1936. This essay originally appeared in the April 1936 issue of *American Forests Magazine* under the title "Threatened Species: A Proposal to the Wildlife Conference for an Inventory of the Needs of Near-Extinct Birds and Animals," pp. 116–119. It has subsequently been republished in different volumes of Leopold's work, one of the most recent being *Aldo Leopold's Southwest*, David E. Brown and Neil B. Carmony, eds. (Albuquerque: University of New Mexico Press, 1995).

10. Charles Jonkel, "Science Under Siege in the Greater Salmon-Selway Bitterroot Ecosystem," a letter of comments written by Jonkel on behalf of the Great Bear Foundation in response to the U.S. Fish and Wildlife Service's decision to adopt the "experimental, non-essential" option as its preferred alternative. It was also posted on the internet by the Alliance for the Wild Rockies, December 13, 2000.

11. Statement issued by Idaho Governor Dirk Kempthorne to the press in late November 2000 in response to the federal government's plan to bring grizzlies back to the Selway-Bitterroot. To get a sense of the immediate local response, read the editorial that appeared in the *Idaho Mountain Express*, the newspaper of Sun Valley, which ran November 29, 2000.

12. "Little Basis In Governor's Opposition to Grizzlies," *Idaho State Journal*, Pocatello, December 26, 2000. From an editorial written by the newspaper's editors. Another editorial of note, "Support Grizzly Recovery Plan," appeared July 10, 1997 in the *Idaho Statesman*, Idaho's largest circulation daily newspaper, based in the capital city of Boise. "Let's hope the approach taken in the preferred plan wins out," the editors wrote. "It is middle-ground and includes citizen management—two key elements that Idahoans and their leaders should support." Yet another appeared in the *Idaho Falls Post-Register* on September 14, 1995. It was titled "A New Day for Species Management?" and states: "The involvement of resource industry representatives, citizens and environmental groups in this plan is exciting. It could set a new trend for the next couple of decades in working out environmental problems and managing the Endangered Species Act for the benefit of all. Imagine the results if we could have had the same cooperation on the spotted owl, the salmon and the wolf."

13. "Grizzly Bear Attacks: Legends and Facts." 1995. A fact sheet published by the National Wildlife Federation. See Stephen Herrero, *Bear Attacks: Their Causes and Avoidance* (New York: Nick Lyons Books, 1985).

14. Personal communication with Ted Kerasote from his research with the National Park Service and Interagency Grizzly Bear Committee for an unpublished column in *Sports Afield*.

15. Scoping document (the precursor to an Environmental Impact Statement) examining options for reintroducing grizzly bears into the Selway-Bitterroot ecosystem, prepared for the U.S. Fish and Wildlife Service by Dr. Christopher Servheen, 1995.

16. From citizen comments submitted to the U.S. Fish and Wildlife Service on its preferred alternative of adopting the citizen's management strategy and managing bears not with the full force of the Endangered Species Act but as an "experimental, non-essential" under the 10-J provision of the Act.

17. Facts derived from ongoing economic analysis of western communities located next to wilderness areas done by Ray Rasker, who heads the Sonoran Institute's Northern Rockies office in Bozeman, Montana. Together

with Dr. Tom Power, economics professor at the University of Montana, Rasker has shown that landscape protection is a vital "quality of life" factor in why some companies choose to start up or relocate in the Northern Rockies. Rasker provided the information in a personal communication and at a conference sponsored by the Foundation for Research on Economics and the Environment in 1999.

18. *This Place on Earth 2001: A Guide to a Sustainable Northwest* (Seattle: Northwest Environment Watch, 2001), p. 6. The statistic on the amount of acreage lost to subdivision was compiled by Dennis Glick at the Greater Yellowstone Coalition, Bozeman, Montana and by The Nature Conservancy, Helena.

19. In 2001, Congressman James Hansen created a special review committee to look at ways the Endangered Species Act could be modified so that it would be less hostile to developers and provide economic incentives for those who choose not to develop property in ways that might harm habitat for threatened species. Together with former Congresswoman Helen Chenoweth-Hage, he has publicly expressed his reservations about restoring grizzlies to the Selway-Bitterroot. I interviewed Congressman Hansen's Communications Director on the House Resources Committee, Marnie Funk, on February 8, 2001 for a story that appeared in the *Christian Science Monitor*, February 16, 2001, "Species Protection Hits Budgetary Wall."

20. *Cadillac Desert* (New York: Viking-Penguin, 1986). In addition to Reisner's book, which is an excellent source that probes how citizens have subsidized water projects in the West, another authoritative source on the general issue of natural resource subsidies in the West is Charles F. Wilkinson, *Crossing the Next Meridian: Land, Water, and the Future of the West* (Washington, D.C.: Island Press, 1992).

21. Daniel Kemmis, "The Reintroduction Question: Bringing grizzlies back to the Bitterroots: Citizen alternative works best," written as a guest editorial published in *The Missoulian* on July 30, 1997. Another piece worth noting was written by former U.S. Forest Service Chief Jack Ward Thomas, a biologist, who today is the Boone and Crockett Professor of Wildlife Conservation at the University of Montana-Missoula. During the late 1980s he was involved in drafting a management plan in the Pacific Northwest to protect the spotted owl. His piece in *The Missoulian*, published October 14, 1997, was titled, "Grizzly Reintroduction: If people aren't involved, the process is misguided." Thomas wrote: "If I were to distill my 40 years as a land and wildlife steward into one significant observation, it would be this: the most significant progress toward conserving our rich natural heritage

has come not from new laws, not through court decisions, and not via groundbreaking scientific research (I say that as a researcher). Rather, such advances happened when people of good will and intelligence stepped back from the rhetorical din that surrounds contentious resource issues and became part of the solution."

Contributors

Ted Kerasote, editor of the Pew Wilderness Center's annual, *Return of the Wild*, has contributed to more than fifty periodicals, including *Audubon, National Geographic Traveler, The New York Times Book Review, Outside,* and *Sports Afield,* where he has written the Environment column since 1987. He is the author of three books—*Navigations; Bloodties;* and *Heart of Home*—and in 1997 won the Wyoming Wildlife Federation's Conservation Communicator of the Year Award. He lives in northwestern Wyoming.

Steven Bouma-Prediger is Associate Professor of Religion at Hope College in Holland, Michigan, and is the author of *Assessing the Ark: A Christian Perspective on Nonhuman Creatures and the Endangered Species Act.*

Vine Deloria, Jr. is a Professor at the University of Colorado, Boulder, a practicing lawyer, and a Lakota Sioux. He has been Executive Director of the National Congress of American Indians and has authored ten books, including *Custer Died for Your Sins* and *God Is Red.* He is recognized as one of today's leading Native-American spokesmen. He lives in Arizona.

Hal Herring's stories and essays have appeared in *The Atlantic Monthly, High Country News, Field & Stream,* and *Bugle,* the journal of the Rocky Mountain Elk Foundation. He lives in western Montana.

Suellen Lowry is the creator and outreach coordinator of Allied Voices, a program that focuses on grassroots partnerships among members of the religious community and secular environmental organizations such as the Earthjustice Legal Defense Fund, the Endangered Species Coalition, and the Biodiversity Project. She has also been a national legislative and PAC director, endangered species lobbyist, congressional staffer, and private attorney. Along with Rabbi Daniel Swartz, she co-authored *Spirituality Outreach Guide, A Guide for Environmental Groups Working with Faith-Based Organizations*.

Chris Madson is the editor of *Wyoming Wildlife* magazine, a monthly conservation periodical that has won more than forty national awards for excellence in writing, photography, and design in the last fifteen years. Madson's writing has appeared in more than a dozen periodicals including *Audubon, National Wildlife*, and *The Nature Conservancy*. The author of *When Nature Heals*. He lives in eastern Wyoming.

Mike Matz has been the associate director of the Northern Alaska Environmental Center, spent time in Juneau as a lobbyist with the Alaska Environmental Lobby, and was then associate field representative for the Sierra Club in Anchorage. The Sierra Club shipped Matz to Washington, D.C. for six years, during which time he was its director of public lands and served as chairman of the Alaska Coalition. He escaped from Washington to lead the Southern Utah Wilderness Alliance in Salt Lake City for seven years. He is now the director of the Pew Wilderness Center.

John McComb is the Pew Wilderness Center's Chief Information Officer. He has also served as an information systems consultant to many major environmental groups including Environmental Defense, The Wilderness Society, Defenders of Wildlife, Southern Utah Wilderness Alliance, Alaska Wilderness League, Mineral Policy Center, and the Sierra Club. He was formerly the Deputy Conservation Director for The Wilderness Society and Conservation Director of the Sierra Club.

Richard Nelson is a nature writer and cultural anthropologist whose work explores the relationships between people and the environment. His books include *Make Prayers to the Raven, Shadow of the Hunter, Heart and Blood: Living with Deer in America*, and *The Island Within*, which won the John Burroughs Award for nature writing. When away from his desk, Nelson's life centers around wildlife watching, surfing, hiking, subsistence hunting and fishing, and exploring the wild coast near his home in southeast Alaska. He is also a conservation activist and public wild-lands advocate, working for protection of old-growth rainforest in the Tongass National Forest.

Thomas Michael Power is Professor of Economics and Chairman of the Economics Department at the University of Montana. He has lived in Montana and taught there for the past thirty-five years. He has written extensively on the economies of the western states and the role played by natural amenities in enhancing economic well-being and transform-ing those economies. His most recent books include *Post-Cowboy Economics: Pay and Prosperity in the New American West* (Island Press, 2001) and *Lost Landscapes and Failed Economies: The Search for a Value of Place* (Island Press, 1996).

Douglas W. Scott has been a wilderness lobbyist with The Wilderness Society and the Sierra Club since 1967, and a leader in the campaign for protection of de facto wilderness in national forests. He was lobbying coordinator for the Alaska Coalition in the enactment of the Alaska National Interest Lands Act, and is currently the policy director of the Pew Wilderness Center.

Michael Soulé is Research Professor (Emeritus) in Environmental Studies, University of California, Santa Cruz, and the founder of the Society for Conservation Biology and The Wildlands Project. He has published more than 150 articles on biological and environmental subjects and was named by *Audubon Magazine* in 1998 as one of the 100 Champions of Conservation of the 20th Century. A Guggenheim fellow, he has co-authored or edited numerous books, including *Reinventing Nature: Responses to Postmodern Deconstruction*. He lives in western Colorado.

Jack Turner taught philosophy at the University of Illinois, was a Woodrow Wilson National Fellow, and has served on the Rhodes Scholarship Committee for the State of Wyoming. He has led over forty-five treks and expeditions in the Himalayas and the Andes, and has climbed in the Teton Range for forty years, often as guide for Exum Guide Service and the School of American Mountaineering. He is the author of *The Abstract*

Wild as well as *Teewinot: A Year in the Teton Range* and has a particular interest in how genetic engineering is domesticating the wild. He lives in northwestern Wyoming.

Todd Wilkinson has been writing about the environment and culture of the American West for two decades. In addition to being a lead environmental correspondent for the *Christian Science Monitor*, his work has appeared in numerous national magazines, among them *Audubon, Orion, Outside, The Utne Reader*, and *High Country News*. He is the author of eight books, including *Track of the Coyote* and *Science Under Siege—The Politicians' War on Nature and Truth*, which describes the struggles of a number of well-known scientific whistleblowers. He has received an Earthwrite Award from the Nike Corporation for an essay about land trusts and conservation easements that appeared in *Big Sky Journal*. He lives in southern Montana.

Florence Williams has written about environmental issues and the West for ten years, first as a staff reporter at *High Country News*. Her articles and essays appear regularly in such publications as *The New York Times, The New Republic*, and *Outside Magazine*. A graduate of the creative writing program at the University of Montana, she received the 1996 and 1998 personal essay awards from the American Society of Journalists and Authors. She currently lives in western Montana.

Index

Page numbers in *italics* indicate illustrations and captions.
Page numbers followed by an *n* indicate endnotes.

RETURN *of the* WILD

About the Pew Wilderness Center

 PEW WILDERNESS CENTER   The Pew Wilderness Center provides economic, historical, and ecological research on the origins and value of wilderness. We assist citizens' groups with their plans and strategies to protect open space, wildlife habitat, and recreational opportunities, and we coordinate educational efforts that highlight the need to safeguard wild places.

The main offices of the Pew Wilderness Center are located in Boulder, Colorado, Washington, D.C., and Seattle, Washington. We do not solicit members. Our intent is to complement rather than compete with other conservation organizations, working with them to protect wild places.

About Island Press

Island Press is the only nonprofit organization in the United States whose principal purpose is the publication of books on environmental issues and natural resource management. We provide solutions-oriented information to professionals, public officials, business and community leaders, and concerned citizens who are shaping responses to environmental problems.

In 2001, Island Press celebrates its seventeenth anniversary as the leading provider of timely and practical books that take a multidisciplinary approach to critical environmental concerns. Our growing list of titles reflects our commitment to bringing the best of an expanding body of literature to the environmental community throughout North America and the world.

Support for Island Press is provided by The Bullitt Foundation, The Mary Flagler Cary Charitable Trust, The Nathan Cummings Foundation, Geraldine R. Dodge Foundation, Doris Duke Charitable Foundation, The Charles Engelhard Foundation, The Ford Foundation, The George Gund Foundation, The Vira I. Heinz Endowment, The William and Flora Hewlett Foundation, W. Alton Jones Foundation, The John D. and Catherine T. MacArthur Foundation, The Andrew W. Mellon Foundation, The Charles Stewart Mott Foundation, The Curtis and Edith Munson Foundation, National Fish and Wildlife Foundation, The New-Land Foundation, Oak Foundation, The Overbrook Foundation, The David and Lucile Packard Foundation, Pew Charitable Trusts, Rockefeller Brothers Fund, The Winslow Foundation, and other generous donors.